Transportation Statistics and Microsimulation

Transportation Statistics and Microsimulation

Clifford H. Spiegelman

Eun Sug Park

Laurence R. Rilett

CRC Press
Taylor & Francis Group
Boca Raton London New York

CRC Press is an imprint of the
Taylor & Francis Group an **informa** business

A CHAPMAN & HALL BOOK

Chapman & Hall/CRC
Taylor & Francis Group
6000 Broken Sound Parkway NW, Suite 300
Boca Raton, FL 33487-2742

First issued in paperback 2022

ISBN 13: 978-1-03-247766-4 (pbk)
ISBN 13: 978-1-4398-0023-2 (hbk)

DOI: 10.1201/9781439894545

Library of Congress Cataloging-in-Publication Data

Spiegelman, Clifford H. (Clifford Henry)
    Transportation statistics and microsimulation / authors, Cliff Spiegelman, Eun Sug Park, Laurence R. Rilett.
        p. cm.
    "A CRC title."
    Includes bibliographical references and index.
    ISBN 978-1-4398-0023-2 (hardcover : alk. paper)
    1. Transportation engineering--Statistical methods. 2. Transportation--Simulation methods. I. Park, Eun Sug. II. Rilett, Laurence R. III. Title.

TA1145.S65 2011
388.01'5195--dc22   2010027449

Visit the Taylor & Francis Web site at
http://www.taylorandfrancis.com

and the CRC Press Web site at
http://www.crcpress.com

Kathy, Lindsey, and Abigail by C.H.S.
Angela, Jeremy, and Eun Gyu by E.S.P.
Katherine, Michael, Samuel, and Beverley by L.R.R.

Table of Contents

Preface

The basic concept of transportation—the movement of goods and people over time and space—has changed little since the Romans developed their transportation system over two thousand years ago. Today we have far more extensive transportation infrastructure systems that include roads, waterways, railways, and air transport options, and much more sophisticated technology than in the past, but the overall objective of transportation engineers remains the same—to plan, design, construct, and maintain the various transportation modal systems in the safest and most efficient manner possible. To achieve this goal, transportation professionals have to be able to answer fairly sophisticated questions, such as:

Which pavement is most economical for a given situation?

What roadway geometry is safer?

What traffic control device works best?

Where should we invest our limited resources to produce the most favorable outcome?

To answer these types of questions, the engineers and planners identify a clear hypothesis, collect relevant data (either through experiment or observation), and develop reasonable conclusions from the data, all of which will require the transportation professional to have a solid grounding in statistics. This text is designed to provide the necessary background knowledge to make informed transportation-related decisions.

With transportation accounting for between 10% and 20% of the U.S. economy, the types of questions listed above are asked thousands of times per day across the country. Unfortunately, textbooks that relate specifically to transportation statistics, to which a transportation professional can turn for help, are very few. Though there are many general engineering statistics books that can provide the necessary background material, these books do not address statistics from the unique perspective of transportation. This textbook helps to fill that gap by discussing statistical concepts in the context of transportation planning and operations.

Our anticipated audience is comprised of transportation professionals, both planners and engineers, who are looking for more sophisticated information than is found in a general undergraduate statistics course, frequently required by a professional program. This textbook would be ideal for an introductory graduate class in transportation statistics that could be taken by first-year students specializing in pavements engineering, transportation systems engineering, or urban planning. In addition, the book would be useful for working professionals who would like to learn more about some of the concepts to which they are exposed during their transportation careers.

While much of what planners and engineers do has been codified (e.g., take twenty random pavement samples and test to failure, etc.), this book will help explain the *why* behind the standard methods. Lastly, we assume the reader has a basic knowledge of introductory probability and statistics as well as a strong working knowledge of basic transportation concepts.

We, the authors, have over fifty combined years of using statistical techniques for transportation research and teaching. Two factors initiated our collaborative development of this textbook: (1) a graduate statistical course established by the authors at Texas A&M University approximately ten years ago, and (2) over ten years of collaboration on various transportation research projects that involved many of the statistical techniques discussed in this book. Based on these experiences, we made two fundamental, yet dichotomous, observations that motivated us to write this text. The first is that many transportation professionals suffer from statistical anxiety—not because the concepts are so difficult, but because they have not been presented in a way that could be easily understood by practicing engineers and planners. We felt the best way to reduce this fear was to use specific transportation examples and problems, all based on real situations using real data, to illustrate the key concepts in the text. The second is that there is also, ironically, a danger of overconfidence. That is, many transportation professions assume that their tried and true statistical methods, which they have become accustomed to using in their professional lives, would continue to be appropriate, even though the underlying assumptions of these statistical techniques are no longer valid. Equally important are the numerous new techniques that could be used in everyday practice but have not been well explained in the available literature. That is, the transportation professionals did not know what they did not know. We hope this textbook fulfills this need.

We have made a concerted effort to define explicitly the underlying assumptions and to provide references and insight so the readers will know when they need to seek outside help from practicing statisticians. In addition, terms have been carefully defined from both a statistical and a transportation perspective. In our experience, the jargon adopted by the transportation and statistics profession often works at cross-purposes; we are hopeful that this text will help make the conversations between transportation professionals and statisticians smoother and more productive for both parties.

KEY FEATURES OF THIS TEXTBOOK

A key feature of this book is that it is focused on realistic transportation-related problems. The sample problems, both within and at the end of the chapters, make use of data obtained as part of various transportation studies around the country. The goal was not to produce a textbook that was exhaustive in nature. Rather, the focus was on statistical techniques most heavily used by pavement and transportation professionals. In particular we discuss:

- The difference among planned experiments and quasi-experiments and field studies

- Strategies for conducting computer-aided statistical designs, fractional factorial designs, and screening designs

- Bias-corrected confidence intervals (we stress that biases in both designing experiments and estimation, unless negligible, should not be ignored)

- Resampling techniques for evaluating uncertainties, including the jackknife and bootstrap, when there is no closed-form estimator for a standard error (this topic is rarely covered in statistics books, but has become increasingly important in transportation studies)

- The concept of Bayesian methods using a conjugate prior approach to Bayesian estimation

- The concept of smoothing estimators in both regression and density estimation.

STATISTICAL SOFTWARE

It is our contention that the reader will best learn the concepts in this textbook by doing—in this case using real data to solve real questions. Each chapter has a number of simple examples so that the readers can familiarize themselves with the underlying theory and equations. However, the state of the practice in statistics is to use statistical programs when solving problems—and the transportation area with its considerable use of data is no exception. As such, we have made a conscious decision to adopt a specific computer software package: JMP by SAS. We chose this package because it is designed to link statistics with graphics so that the user can interactively explore, understand, and visualize data, which we have found useful when solving transportation statistical problems. In fact, one of our first chapters is dedicated to this field. It also has both Windows-based and Macintosh operating systems, which our students have found useful. Please note that the techniques in this book are, of course, applicable to many of the fine statistical packages available on the open market, such as SPSS, BMDP, and SYSTAT, among others. While we could have chosen to write the textbook in a general format without any reference to computer packages, we felt it would not be as useful to the reader. Lastly, we are confident that an informed reader could take the lessons from this textbook and successfully apply them using any statistical package.

In addition, we provide the opportunity for student readers to buy their own copy of JMP at a significantly reduced price. This will allow students at universities that do not have a JMP site license the opportunity to design and analyze experiments of their choosing.

ORGANIZATION OF THE TEXTBOOK

The textbook is comprised of 15 chapters. The first chapters are related to standard probability and statistical techniques that the user will need for the more advanced sections of the book. These are included in Chapters 2 through 5 and cover the basics of graphical methods (which are emphasized throughout the book), numerical summary methods, random variables, probability mass functions, and probability distribution functions.

Chapters 6 to 9 introduce sampling distributions as well as techniques for comparing observed results with various hypotheses about what those results should or could be. These latter techniques are known as statistical inferences, and we cover single and multiple variables as well as continuous and categorical data. Chapter 10 and 11 are related to model building with a particular emphasis on regression. The reader will be exposed to both simple and multiple linear regression for continuous data and generalized linear models for count data. Chapters 12 to 14 are related to experimental design, uncertainty estimation, and Bayesian estimation. In addition, new methods for identifying standard errors, when the assumptions for commonly used approaches are not applicable, are introduced. An introduction to Bayesian approaches is provided and is based on the use of conjugate priors. Lastly, Chapter 15 deals with the issue of traffic microsimulation models and how the techniques developed in the preceding chapters can be used to test various hypotheses.

A draft version of this textbook was used to teach a transportation statistics course at Texas A&M University to transportation and materials graduate students. A number of topics have been added since the course was taught. We assume that a week of lectures is 150 minutes long. The list of time that it took to cover each covered topic is given in the table below.

Chapter	Time in Weeks
2	0.75
3	1
4	1
5	1
6	1
7	1.5
8	1.5
9	1.5
10	1.5
11	1
12	1.5
13	0.5
14	1
15	1.5
Appendix	0.5

Acknowledgments

This textbook was a collaboration in the true sense of the word—and would not have been accomplished without the contributions of many of our colleagues and institutions. First, we acknowledge the students who have taken our graduate courses over the years. Many of the problems and much of the text were "beta tested" on them, and we are truly appreciative of the feedback they provided. The text itself was edited numerous times, and we acknowledge the excellent work of our primary editor, Lindsay Mayo-Fincher, and the editorial assistance of Vanessa Steinroetter, Lisa Dworak, and Beverley Rilett.

One of the strengths of the text is in the use of transportation data that was collected from various research projects at Texas A&M University and the University of Nebraska. We express our sincere gratitude to Paul Carlson, Bill Eisele, David Ellis, Kay Fitzpatrick, Tom Freeman, Tim Lomax, Eyad Masad, Jennifer Ogle, David Schrank, Shawn Turner, Jerry Ulman, Justice Appiah, Bhaven Naik, and Cesar Quiroga for providing data used in this textbook.

We are very appreciative for the encouragement that we received from our colleagues and the administration at Texas A&M University, the Texas Transportation Institute, the University of Nebraska–Lincoln, and the Nebraska Transportation Center. We also gratefully acknowledge the financial support that allowed this work to be accomplished. In particular, the following organizations provided grant funding for portions of our textbook:

- U.S. Department of Transportation, University Transportation Centers Program to the Southwest Region University Transportation

Center (it is also funded, in part, with general revenues from the State of Texas), and

- U.S. Department of Transportation, University Transportation Centers Program to the Mid-America Transportation Center, which is headquartered at the University of Nebraska–Lincoln.

- We thank Abby Spiegelman for a prototype sketch of the textbook cover and Bethany Carlson for creating the final cover design.

About the Authors

Dr. Clifford Spiegelman is a distinguished professor of statistics at Texas A&M University, where he has been for twenty-three years. Dr. Spiegelman has also been a senior research scientist at Texas Transportation Institute (TTI) for about fifteen years. He held a position at the National Bureau of Standards (now NIST). Dr. Spiegelman has been a member of TRB statistics committees for many years, and is a founding member of the American Statistical Association transportation statistics interest group.

Dr. Eun Sug Park is a research scientist at TTI, where she has worked for the past nine years. Prior to joining TTI, she was a research associate at the University of Washington's National Research Center for Statistics and the Environment. An author or coauthor of numerous papers and research reports, Dr. Park has been honored with the TRB Pedestrian Committee Outstanding Paper Award (2006 and 2009) and the 2009 Patricia Waller Award. She holds a PhD in statistics from Texas A&M University, an MS in statistics, and a BS in computer science and statistics, both from Seoul National University.

Dr. Laurence R. Rilett is a distinguished professor of civil engineering at the University of Nebraska–Lincoln. He also serves as the director of both the U.S. Department of Transportation's Region VII University Transportation Center (the Mid-America Transportation Center) and the Nebraska Transportation Center. Dr. Rilett received his BASc degree and his MASc degree from the University of Waterloo, and his PhD degree from Queen's University. He has held academic positions at the University of Alberta and Texas A&M University. Dr. Rilett is an associate editor of the ASCE *Journal of Transportation Engineering* and is on the editorial board of the *Journal of Intelligent Transportation Systems: Technology, Planning, and Operations.*

How to Contact the Authors and Access the Data Sets

The writing of this textbook was a much greater challenge than the authors had originally thought—after all, we had been working collaboratively for over ten years and had cotaught a number of statistical courses for transportation engineers. After writing the text and working through our material as carefully as possible, we acknowledge that there are sure to be some unanticipated glitches that we missed, that transportation examples exist that we might have used, that other topics should have been included, and that we could have identified more challenging example problems. We welcome readers' comments, suggestions, and real transportation data that may be used as example problems in further editions of this text. We can be reached at http://www.stat.tamu.edu/transstat. Most of the data used in this book are available in both JMP and tab-separated format and can be found on this web site as well.

CHAPTER 1

Overview

The Role of Statistics in Transportation Engineering

MOST TRANSPORTATION PROFESSIONALS WILL take an introductory course in probability and statistics as undergraduates and, if they pursue further studies, a more in-depth statistics course as graduate students. Rarely do these courses have a specific transportation focus. This is problematic because the problems that they will encounter in their working careers tend to be different from those taught in the generic introductory courses. The relationships that need to be modeled are complex, and the variables are often categorical in nature. In addition, a large proportion of transportation studies are observational in nature and are not amenable to experimental design. Accordingly, the commonly used statistics in transportation reflect the unique characteristics of its subjects and are often not covered in detail in general statistics classes. Because statistics have become so prevalent in the practice of transportation engineering, a statistics textbook written with a transportation perspective and employing transportation-related problems is essential to the profession.

It is the authors' contention that because the transportation problems that engineers address are becoming more complex, the statistical knowledge base of transportation engineers needs to be developed in such a way that they will (1) have a deeper appreciation of the techniques they use in their day-to-day working environment, (2) know when to bring in statisticians to help them with complex problems, and (3) be able to communicate in a meaningful way to the statisticians with whom they will be working. While it is clearly impossible to discuss every statistical technique in a single text-book, this book does focus on those most likely to be encountered by trans-portation professionals. In order to expand on this point further, it is helpful to understand the working environment of the transportation professional.

1.1 WHAT IS ENGINEERING?

The word *engineer* originates from the Latin term *ingenerare*, meaning "to invent, to create, or to regulate." The Accreditation Board for Engineering and Technology (ABET) defines engineering as

> the profession in which a knowledge of the mathematical and nat-ural sciences gained by study, experience, and practice is applied with judgment to develop ways to utilize, economically, the materi-als and forces of nature for the benefit of mankind. (ABET, 1986)

The layperson is often confused by the roles of scientists and engineers. This is probably because both the engineer and the scientist are thoroughly educated in the mathematical and natural sciences. The main difference is that scientists primarily use this knowledge to acquire new knowledge, while engineers apply the knowledge to the design and development of usable devices, structures, and processes. This difference has been summarized— and no doubt taught to numerous introductory engineering classes—as "the scientist seeks to know; the engineer aims to do" (Eide et al., 1979).

Engineering in its simplest form is problem solving. To this aim engineers develop devices, processes, structures, and systems based on the application of a detailed knowledge of science (e.g., physics, chem-istry, material behavior) and of mathematics (e.g., integral calculus, differential equations, statistics) in an affordable and efficient manner. Similar to economists, engineers need to understand the relationship between *demand* and *supply*, though these terms have different mean-ings from the same terms as used by an economist. As an example, to a transportation engineer the demand may be the number and types of vehicles wanting to use a roadway, while the supply is the components that comprise the transportation system (e.g., road width, number of lanes, design standard, etc.). The transportation engineer seeks to find the best arrangement of road components (e.g., the supply) that meets the various users' needs (e.g., the demand) and does so in an economical manner. Note that the demand and supply are not deterministic; thereby, engineers must utilize the concepts of probability and statistics in their everyday working environment (Ang and Tang, 2006; Benjamin and Cornell, 1970).

In many ways the problems faced within the profession are similar to those faced by engineers throughout recorded history. For example, the question of how to move freight from one point to another across vast distances and over diverse geographic regions has been a problem that has attracted engineering attention for thousands of years. Approximately 175 years ago a popular method to solve this problem was the use of canals. Arguably, the most famous canal in the United States was the Erie Canal, which links the Hudson River (and hence New York City) with Buffalo (and thereby the western United States via the Mississippi waterway via Chicago via the Great Lakes), while simultaneously bypassing Niagara Falls, which was a major barrier to movement by ship. Figure 1.1 shows a picture of a barge on an Erie Canal aqueduct in Rochester, New York. It can be seen that if the engineers wished to move freight into and out of the city, they had to first cross the Genese River. Their solution was to build an aqueduct across the river—in effect a grade separation between the canal and the river. The aqueduct was originally built in 1825 and rebuilt in 1842; indeed, the aqueduct still stands today, although it is now used as a

FIGURE 1.1 Erie Canal aqueduct over the Genesse River in Rochester, New York. (From the collection of the Rochester Public Library Local History Division. With permission.)

roadway. At the time, the construction of the Broad Street Aqueduct was a phenomenal feat of engineering, as is evidenced by Marquis de Lafayette's proclamation during his tour of the United States in 1825:

> The grand objects of nature, which threatened to impede, have been made only to adorn, as we see in the striking spectacle which is at this moment presented to our enchanted eye. I enjoy the sight of works and improvements equally rapid and wonderful—among which is this grand canal, an admirable work, which genius, science, and patriotism have united to construct.

Freight can be moved a variety of ways today—in containers on trucks, ships, airplanes, and railway cars. It still, of course, is moved via canals. Figure 1.2 shows a portion of a 918-meter aqueduct, known as the Wasserstrassenkreuz Magdeburg, over the Elbe River at Magdeburg, Germany. The aqueduct was opened in 2003 and connects the Midland Canal and the Elbe-Havel Canal.

A comparison of Figures 1.1 and 1.2 shows that while the technology used by engineers can change, the issues faced by engineers, such as how to move freight, how to move people, and how to design the infrastructure, do not. What has changed is that we have a greater understanding of the uncertain and stochastic nature of the various variables that affect supply

FIGURE 1.2 Aqueduct over the Elbe River at Magdeburg, Germany. (From AP/Wide World Photos. With permission.)

and demand. As a result, all facets of transportation and engineering make extensive uses of statistics—and without a doubt, the engineers of the Magdeburg Aqueduct did so as well. Understanding the best technologies to use, how to use them, and when to use them is not always obvious. This requires research, study, and to a great degree statistical analysis, which is the underlying motivation of this textbook.

1.2 WHAT IS TRANSPORTATION ENGINEERING?

Civil engineering is one of the major branches of engineering and, as its name implies, is related to engineering for civilian applications. Civil engineering improves quality of life through the provision of infrastructure such as:

- Buildings, bridges, and other structures

- Highways

- Dams and levees

- Water treatment and waste disposal plants

Transportation engineering is a branch of civil engineering that is involved in the planning, design, operation, and maintenance of safe and efficient transportation systems. These systems include roadways, railways, waterways, and intermodal operations. Typically, the demand is the amount of traffic (people, cars, railcars, barges) that is expected to use a particular transportation facility, while the supply is the quantity and type of infrastructure components (roadways, bridges, pavements, etc.). These systems are typically large and expensive.

There are a number of attributes of transportation engineering that affect the types of statistical theory that are used in the profession. One important aspect of transportation engineering is that the transportation engineer is not only interested in the infrastructure (e.g., bridges, rails, etc.) and the individual units (cars, trucks, railcars) that use the infrastructure, but also the user. A conceptualization of this environment is shown in Figure 1.3. Often it is necessary to understand the interaction of all three of these entities—infrastructure, individual units, and user—to understand the system as a whole. Typically the infrastructure and units are considered the supply side of the equation, while the users are identified with demand.

Experimental studies, or designed experiments, are the mainstay of many standard statistics books. They are used extensively in many engineering

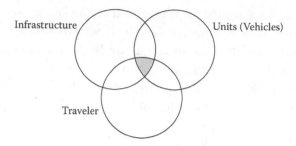

FIGURE 1.3　Conceptualization of the transportation environment.

disciplines, including pavement engineering, that are not necessarily applicable to transportation systems engineering. For example, consider an engineer who is interested in the various factors that affect skid resistance and the relationship to crash rate. From an ethical standpoint, she cannot place various types of pavement surfacing on different sections of highway, observe what types of accidents occur, and then choose the best type of pavement based on the empirical accident results. Instead, most transportation studies are observational in nature, and as a result, the statistics used by transportation engineers reflect this characteristic. In addition, it is sometimes very difficult to obtain certain data from the transportation system, so statistical techniques that can handle missing data or use *a priori* knowledge are needed. Lastly, much of the data are correlated and interdependent. For example, the travel time on a given link is often correlated to the travel time on the immediate downstream link. Sometimes this correlation is negative: Consider, for example, a driver stopped at a traffic signal that is red. If the signal system is coordinated properly, the driver will have a lower probability of being stopped at the traffic signal on the next link. At other times, however, the correlation is positive: If one link is experiencing high travel times because of excessive demand, then other links also will experience high travel times because of the same demand. Regardless, as this example demonstrates, the assumption that different transportation phenomena are independent is not always valid.

In the United States transportation is estimated as representing 10% of the nation's gross domestic product (USDOT, 2009), and a well-maintained and comprehensive system is considered by many to be a necessary condition for a successful economy. However, the American Society of Civil Engineers (ASCE) in their annual report card gave a grade of D– to the U.S. roadway system. As the report notes, "One-third of America's major roads are in poor or mediocre condition and 45 percent of major urban highways are congested. Current spending of $70.3 billion per year for highway capital improvements is well below the estimated $186 billion needed annually to substantially improve conditions." Given the environment of inadequate funding and critical maintenance needs, it is crucial that transportation engineers be able to understand the demands (e.g., goods and people movement) on the system as well as the characteristics of the supply (pavement, roadway geometrics, etc.). It is the premise of this book that a knowledge of statistics is critical to understanding the trade-offs so that our limited resources can be used as effectively and as efficiently as possible.

1.3 GOAL OF THE TEXTBOOK

The goal of this textbook is to introduce transportation students to important statistical techniques that are commonly used in transportation engineering. The objective is not to produce an engineer who can replace the need for a statistician. Clearly, an engineer who desires the depth and breadth of statistical knowledge required to do this would need to complete an advanced degree from a university statistics department. Instead, the goal of this textbook is to provide (1) a strong statistical background for transportation professionals that enables them to perform their various job functions better, and (2) the necessary knowledge so that transportation engineers can communicate effectively with statisticians who can help address the complex issues faced by many engineers. It is the authors' belief that the transportation problems have become so complex that only a truly interdisciplinary approach will be successful. As this approach requires interaction and a basic foundation of knowledge, this book is designed to facilitate this essential dialogue between statisticians and engineers.

1.4 OVERVIEW OF THE TEXTBOOK

Comprised of fifteen chapters, this textbook can be divided into five main sections. The first section of the textbook explores the conventional, but essential, means of interpreting data, including graphical analysis (Chapter 2) and common summary measures (Chapter 3). The next section includes basic definitions of probability and random variables (Chapter 4) and common probability distributions found in transportation engineering (Chapter 5). The third section is related to basic statistics, including sampling distributions (Chapter 6), statistical inference (Chapter 7), and ANOVA and distribution-free tests (Chapter 8). The fourth section includes categorical data analysis (Chapter 9), regression techniques, including linear regression (Chapter 10), and regression models with discrete dependent variables (Chapter 11). The fifth section includes experimental design; factorial, fractional factorial, and screening designs; D-optimal and I-optimal designs; sample size calculations; and field and quasi-experiments (Chapter 12). Furthermore, the cross-validation, jackknife, and bootstrap methods for obtaining standard errors are also included in this section (Chapter 13). The textbook ends with two chapters that are not often treated in standard statistical books, but are becoming increasingly important to transportation professionals. The first covers Bayesian approaches to transportation data analysis (Chapter 14), and the second addresses statistical methods for microsimulation models (Chapter 15).

1.5 WHO IS THE AUDIENCE FOR THIS TEXTBOOK?

This textbook is designed for an introductory graduate or upper-level undergraduate class in transportation statistics. It is anticipated that the students will have a working knowledge of probability and statistics from their undergraduate course or an introductory graduate-level course. Terms such as *random variable*, *probability density function*, *t-test*, and other statistical vocabulary should be familiar to students using this resource. The book examines current statistical techniques for analyzing transportation problems and uses actual transportation data and problems to illustrate the procedures. Yet, the audience for this textbook is certainly not limited to university students; we believe that the textbook also will be useful to the practicing transportation professional. The examples culled from the current state of the practice will help working professionals obtain a better grasp of the theory and issues behind the different techniques commonly used in the profession today. Thus, though the book is written from an engineering perspective, it will be useful to anyone working in the transportation field (e.g., urban planners, management majors, etc.).

1.6 RELAX—EVERYTHING IS FINE

We have taught transportation statistics for a combined total of more than fifty years and have experienced many instances of engineering students' anxiety about using statistical analysis. Students often have a palpable fear associated with taking a statistics course, which we attribute partly to unfamiliar terminology. Statisticians, similar to engineers, have developed their own language, and this can be daunting to someone who is not trained in the discipline. We have endeavored to write all our explanations in a clear and simple style. Because not every topic can be treated in the detail necessary, we have endeavored to add appropriate references throughout the text. Another reason for student anxiety in relation to statistics is that many courses do not use scientific or engineering examples, much less transportation examples. Moving from fairly generic concepts to particular transportation problems can be daunting. For this reason, we have used real transportation problems throughout and have attempted to list the pitfalls of the different techniques. The authors also have made available a discounted version of JMP, a statistical software package developed by the SAS Corporation, with the textbook so that readers can try the presented examples on their own. While the textbook examples are mainly based on JMP, the techniques described in the textbook can be used by any standard statistical package.

In sum, this book is geared toward engineers who want to develop a working knowledge of statistics and desire material that is conveyed in a simple and straightforward manner. While we know that it is not possible to cover all statistical areas in depth, we have tried to demonstrate the intuitive nature of statistics and hope that this will make it more comprehensible for the reader. With this aim in mind, the goal of this textbook is to function as a supplement to statistics that has transportation as its nexus. We hope that a student, worker, or professional in the transportation field who uses this textbook will be able to engage in an effective dialogue with statisticians. For readers who are apprehensive about statistics, all we can do is give the advice we give our own students routinely: "Relax—everything will be fine."

REFERENCES

Accreditation board for engineering and technology, 1985. 1986. Annual Report. New York: Accreditation Board for Engineering and Technology.

Ang, A. H. -S., and W. H. Tang. 2006. *Probability concepts in engineering planning and design: Basic Principles.* Vol. 1. New York: John Wiley & Sons.

Benjamin, Jack, R., and C.A. Cornell. 1970. *Probability, statistics and decisions for civil engineers.* New York: McGraw-Hill.

Eide, A. R., R. D. Jenison, L. H. Mashaw, and L. L. Northrup. 1979. *Engineering fundamentals and problem solving.* New York: McGraw Hill.

Rochester Telegraph, June 14, 1825, No. 51-Volume VII, U.S. Department of Transportation (USDOT), Research and Innovative Technology Administration, Bureau of Transportation Statistics. 2009. National transportation statistics. http://www.bts.gov/publications/national_ transportation_statistics/

CHAPTER 2

Graphical Methods for Displaying Data

2.1 INTRODUCTION

As stated in Chapter 1, the objective of this textbook is to help transportation engineers—who are in practice, education, or research—learn basic statistical techniques. The first step in any engineering study should be a graphical analysis, and in this text we use the term *graphical* liberally. For example, consider the map of current speeds from the city of Houston's traffic management center, TRANSTAR, shown in Figure 2.1.

On this map, the darker the freeway link is shaded, the more severe the congestion is on that link. The darkest shaded links indicate a speed of less than 20 miles per hour. As this image illustrates, there is severe congestion on the ring road and on the roads entering the city. Conversely, the roads on the outskirts of the city are relatively clear, and from this information, we can deduce that this is probably a rush hour.

This raw data were obtained from multiple Texas Department of Transportation (TxDOT) Automatic Vehicle Identification (AVI) sensors located on the traffic network. Data were then compiled by the Houston Traffic Management Center, quantified, and displayed graphically for the public to use. Maps of this type are clearly much more informative to the traveling public than listing the links, locations of the links, and current speeds in a table, as shown in Table 2.1.

However, for an engineer who wants the information for other uses, this latter table may be more appropriate. This chapter will introduce the

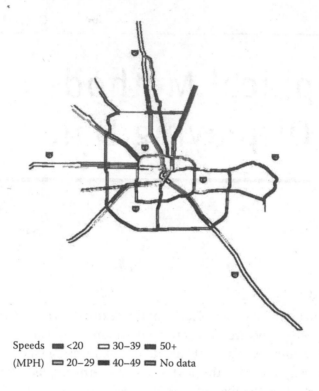

Speeds ▬ <20   ▭ 30–39   ▬ 50+
(MPH) ▭ 20–29   ▬ 40–49   ▭ No data

FIGURE 2.1   Houston speed map. (From Texas Department of Transportation. With permission.)

TABLE 2.1   Travel Time and Speed Table from Houston, Texas

From	To	Distance (miles)	Travel Time (m:s)	Speed (mph)
Ella	IH-10 Katy	2.10	7:30	16
IH-10 Katy	Westheimer	3	7:40	23
Westheimer	US-59 Southwest	2.10	3:07	59
US-59 Southwest	Evergreen	1.40	1:24	60
Evergreen	South Post Oak	1	0:54	66

most common types of graphical techniques used in transportation engineering. It is always wise not only to graph the data once, but also to do so in a number of different ways in order to attain a greater knowledge of the data.

Good plots are crucial to gaining information from any data set. This is particularly true when either the sample size or the number of variables in a data set is large. The common idiom "a picture is worth a thousand words" is applicable; namely, a few good plots can lead to more information than the application of many tests and formulas. Well-chosen plots can help us understand the relationships that underlie the data.

In this chapter we present plotting procedures that have long been used by statisticians and transportation professionals. Histograms and scatter plots are examples of such plotting procedures. Other plots that we present are used by the statistical community, but are not widely used in transportation, e.g., parallel coordinate plots and control charts. Finally, we urge creativity in plotting, as the goal of plotting should be to communicate a deeper understanding of data and not to limit ourselves to a fixed set of tools. Examples of this approach include transforming the variables in our data set to a more meaningful scale before we plot. Readers who want a comprehensive reference for exploring data graphically should see Tukey (1977). Additional reading on this topic includes Leinhardt and Leinhardt (1980) and Theus and Urbanek (2008). It is a good idea to use computer software to generate the graphs. The graphs in this chapter were generated using the JMP statistical software package (SAS Institute, 2009). Step-by-step algorithms are given so that the user better understands how these plots are constructed.

2.2 HISTOGRAM

The histogram shows the relationship between the levels of a variable and the relative frequency of its values. An example of a frequency histogram is shown in Figure 2.2 for Houston Automatic Vehicle Identification

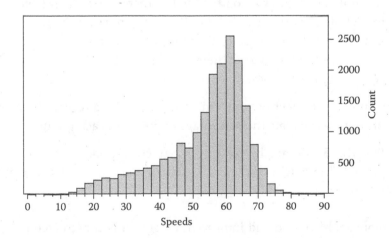

FIGURE 2.2 Histogram of highway speeds in Houston (mph).

(AVI) speed data. We notice rather quickly that a typical speed is around 60 miles per hour. Notwithstanding, speeds are considerably dispersed because there are a number of very slow speeds, too. Slow speeds may indicate incidents or accidents, and thus we gain more insight from the histogram than we would from a table. In addition, this type of information is often required by U.S. transportation authorities. For example, in order for states to receive federal funding, one of the requirements is to provide speed distribution data from selected locations and times throughout their network. If it was found that a given highway had excessive speeding, then funds may be withheld until the problem is corrected—typically through tighter enforcement. For this reason, many state DOTs have permanent staff dedicated to collecting and archiving speed data (among other data) throughout their networks in order to meet federal reporting requirements. A first step in this process would be to plot a histogram to give a visual clue whether there is excessive speeding, and if so, to what extent. The graphical analysis for these data continues well beyond the histogram, but first we give the steps needed to construct a histogram.

Steps to construct a frequency (or relative frequency) histogram are as follows:

1. Sort the data from minimum to maximum.

2. Divide the sorted data into k equally spaced groups. It is best to choose k so that no data fall on the group boundaries. Note that JMP chooses k automatically. Readers can make their own choice using the grabber (hand) tool on the toolbar menu. (See the JMP manual for detailed instructions for modifying the number of groups.) The minimum value belongs to the first group and the maximum value belongs to the kth group. There should be no spaces between adjacent group boundaries.

3. Count the number of points in each group. For a relative frequency histogram calculate the percentage of points in each group.

4. Plot the histogram by plotting a bar corresponding to each of k groups. The height of the bar is proportional to the frequency (or relative frequency) of each group.

A more sophisticated and improved histogram is constructed by plotting a fitted probability density function (pdf) that shows the relationship between the histogram and an assumed model. For example, one might believe that the speeds come from a Gaussian distribution (which is also referred to as normal or bell curve distribution) with a mean μ and a standard deviation σ (see Chapter 5 for more details related to this curve). In this case, one would calculate the sample mean, $\bar{x}$, and sample variance, s^2, from the data and substitute these for the population mean, μ, and the population variance, σ^2, respectively, in the formulas for the normal probability density function:

$$f(x\,|\,\mu,\sigma)=\frac{1}{\sqrt{2\pi}\sigma}\exp\left(-\frac{1}{2}\left(\frac{x-\mu}{\sigma}\right)^2\right).$$

Additionally, it is often helpful to overlay a model-free or smoothed histogram. By adding the smoothed nonparametric histogram, or density, it is possible to get an idea of the number of modes or bumps in the probability model underlying the data. The steps for constructing a smoothed density estimate are discussed in the appendix.

To illustrate the above concepts, Figure 2.3 shows the same speed data as in Figure 2.2 with an overlaid fitted normal curve and the smoothed

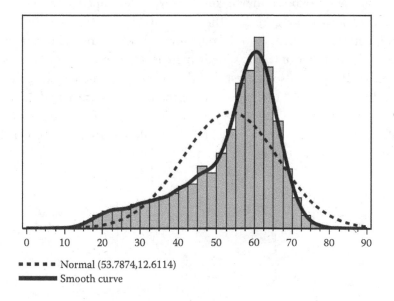

■ ■ ■ ■ ■ Normal (53.7874,12.6114)
━━━━━ Smooth curve

FIGURE 2.3 Normal curve and nonparametric density curve overlaying the histogram of the highway speed data.

nonparametric density curve. The normal density curve is dotted and the smoothed density is solid. The nonparametric density estimate is much smoother than the histogram and may do a better job of showing the nature of the probability model underlying the data. We will discuss how we can test statistically whether a given set of data follow a particular pdf in Chapter 9.

2.3 BOX AND WHISKER PLOT

The next plot that we consider is the box and whisker plot. This plot gives a quick view of many essential features of a data set; in particular, the box and whisker plot is useful for identifying the median, interquartile range (the spread of the middle half of the data), and outliers. They are particularly effective plots to use when comparing different populations.

For example, we may wish to explore the relationship among average speeds as derived from loops and the number of traffic signals along a roadway. Suppose that time mean speed data were collected from different roadways of equal length, but with varying levels of traffic signals. In this scenario assume that the link volumes are approximately equal and the links are located in similar types of urban environments. In order to see whether there is any relationship between median speed and arterial spacing, a series of box and whisker plots were developed for each configuration, as shown in Figure 2.4. The horizontal line within each box of the plot illustrates the median of the data for each scenario, and it can be seen that as the number of traffic signals increases, the median speed tends to decrease. The plots also suggest that this decrease in median speed occurs at an increasing rate. The upper end of the box is an estimated 75% point (75th percentile; see Chapter 3) and the lower end of the box is the estimated 25% point (25th percentile) for each data set. The lines, or whiskers, extending from the ends of the boxes go to the farthest data point within

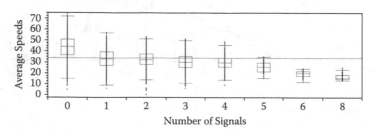

FIGURE 2.4 Box and whisker plots of average travel speeds (mph) vs. number of signals.

1.5 interquartile ranges of the end of the box. The width of each box being one interquartile range (see Chapter 3) and any points graphed beyond the whiskers are known as outliers. Figure 2.4 shows that as the number of signals increases, the median speed decreases. We can also see that the speed decrease is not linear in the number of signals. As the number of signals increases, the variance in average travel speeds tends to decrease to the point that when there are six signals in a segment, there is not much variation in the average speed that drivers use.

The specific algorithm for constructing box and whisker plots is as follows:

1. Calculate the 25th percentile, the median, and the 75th percentile from the data (see Chapter 3).

2. Begin by creating the rectangle of the box and whisker plot. Construct a rectangle that has a lower edge as the 25th percentile and an upper edge as the 75th percentile. A line is drawn parallel to the lower and upper edges of the rectangle at the median. The box spans the middle half of the data.

3. Next, compute the interquartile range. The interquartile range is the difference between the 75th percentile and the 25th percentile (see Chapter 3). It should be noted that a common, "back of the envelope" estimate for the population standard deviation is 3/4 times the inter-quartile range. Thus, 1.5 interquartile ranges are approximately two standard deviations, and 3 interquartile ranges are approximately four standard deviations.

4. Then draw the whiskers. From both ends of the box draw a line to the farthest observation that still lies within 1.5 interquartile ranges from the closest end of the box. Note that it is possible that there are no such observations, and in that case, the line(s) is not drawn.

5. Finally, some programs denote outliers on the box and whisker plot. Any point greater than 1.5 interquartile ranges from an edge of the box is a mild outlier and is plotted with an open circle. Any point beyond three interquartile ranges from an edge of the box is a severe outlier, and some programs use special symbols to plot outliers in this range. For example, mild outliers can be represented by open circles and extreme outliers can be represented by closed circles.

6. The observed points can be added to the graph.

2.4 QUANTILE PLOT

The third type of plot that we consider is a q-q or quantile plot. It is used to assess the assumption of an underlying probability distribution model for the data. Frequently the model that we wish to refer to will be the normal or Gaussian model, but the plotting method works for any continuous probability model. Usually on the y-axis the theoretical percent points are plotted for a standardized form of the distribution, and on the x-axis we plot the observed data. The steps for producing a q-q plot are:

1. Sort the data from smallest to largest observations. Denote the sorted observations as $x_{(1)} \leq x_{(2)} \leq \ldots \leq x_{(n)}$.

2. We compute the $i/(n + 1)^{\text{th}}$ percentiles from our chosen distribution, and denote these as $F^{-1}(i/(n + 1))$. (Refer to Chapter 4 if you are unfamiliar with this notation.)

3. We construct the q-q plot by plotting the pairs $x_{(i)}, F^{-1}(i/(n + 1))$.

4. Some programs, like JMP, add additional features, such as 95% confidence intervals and a probability axis.

Figure 2.5 is a q-q plot of average speeds obtained from an inductance loop detector located on Interstate 37 in San Antonio, Texas. The data were

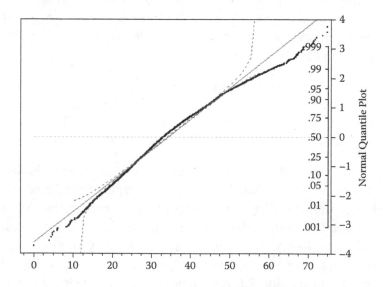

FIGURE 2.5   Quantile plot (q-q plot) of average speeds obtained from inductance loop located on I-37 in San Antonio.

obtained from Figure 2.4 and include all the data from that figure (e.g., the number of signals is not considered in this plot). If the normal model were close to a perfect description for the average speed data, the fitted straight line would run through the plotted points. The q-q plot shows a reasonably good agreement with a normal distribution. We can see that almost all the quantiles lie within the 95% confidence intervals. Keeping in mind that no model is exactly correct, it would be reasonable to assume that these data are adequately modeled by the normal distribution. We will discuss this issue in greater detail in Chapter 9.

Throughout the twentieth century these types of analyses were performed by engineers using specialized graph paper, appropriately named probability paper. With the advent of sophisticated micro-computer-based statistical software, such as JMP, this manual approach is not as widely used.

2.5 SCATTER PLOT

The next type of plot that we consider is the scatter plot. It is used to show the relationship between two variables (X and Y) that are typically modeled as having a continuous scale. Scatter plots are used to assess the degree of correlation and suggest predictive formulas between two variables. Say that we have the bivariate data set consisting of n pairs of measurements (numbers) on X and Y variables, that is, (X_i, Y_i) $(i = 1,...,n)$. The steps for constructing a scatter plot are given below:

1. Denote each observation (X_i, Y_i) by a point on a rectangular coordinate system with the horizontal axis spanning the range of X values and the vertical axis spanning the range of Y values.

2. If desired, plot a least squares line or other model through the plot. This is an option in JMP.

Figure 2.6 shows the scatter plot of population density vs. freeway miles traveled for several large U.S. urban areas. We overlaid a fitted straight and a nonparametric or model-free curve produced by using a smoothing spline on the plot (see appendix at the end of the book). We can see a high degree of negative correlation between these two fitted curves. While the straight line looks to be an adequate fit ($R^2 = .93$; see Chapter 10), the smoothing spline suggests that the true relationship may not be linear. Most of the time, a straight line is more than

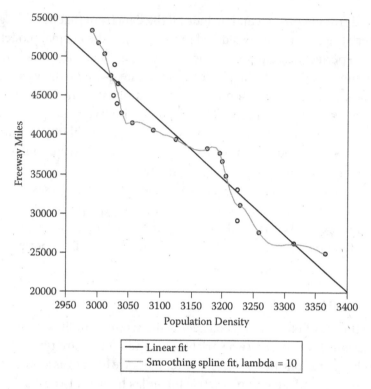

FIGURE 2.6 Population density vs. freeway miles for very large urban areas.

adequate for data similar to this, and in transportation, a linear relationship is the most common simplifying model. However, it is always important to check that this assumption is reasonable by first plotting the data.

When we have several continuous variables that we wish to explore together, we can use a scatter plot matrix. It is a collection of scattered plots combined in a matrix so that the relationships among all the variables can be explored at one time. It is usually a good idea not to plot more than five variables at a time using a plot matrix; otherwise, the plots overwhelm the user. This happens in part because each of the plots must be reduced in size in order to be accommodated on a single page.

An example may be seen in Figure 2.7, where population density, annual public transportation usage in millions of miles, and annual hours of delay in thousands of hours are plotted for twenty-two large urban areas. While a bit overwhelming, it is possible to make general conclusions about the data that would not be as easy if one were looking at it in raw or tabulated form. As an example, we can see from Figure 2.7 that there is a negative

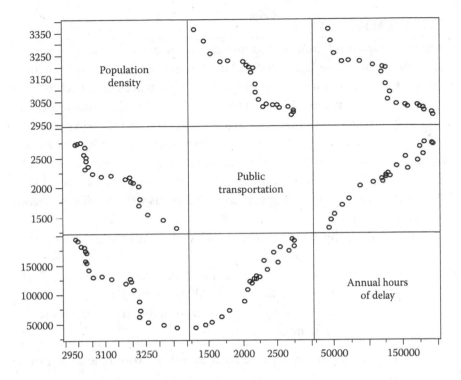

FIGURE 2.7 Scatter plot matrix of population density, annual passenger miles on public transportation, and annual hours of delay.

correlation between population density and annual passenger miles on public transportation, and a positive correlation between annual hours of delay and public transportation usage. In high-density areas there may be shorter commutes, and when there are long delays on highways, there is an incentive to use public transportation.

The old saying "correlation is not caution" should always be kept in mind when interpreting data. Just because two variables are correlated does not mean that one causes the other or, as is more common, that changing one variable will change the other. Frequently, there is a hidden variable that has a causal relationship with both variables. For example, it can readily be shown that highways consisting of thicker pavements sometimes require repair sooner than highways consisting of thinner pavements. Even if there is a negative correlation between pavement life and thickness, it would be erroneous to advocate using thinner pavements on highways. The reality is that the highways that have thicker pavements also tend to have higher volumes, and consequently, this explains the apparently counterintuitive correlation.

2.6 PARALLEL PLOT

Another important tool for examining the relationship among several variables at one time is a parallel plot. Consider Figure 2.8. It is a plot of the passenger miles (on the y-axis in millions) traveled in each of four metropolitan areas (on the x-axis) from 1982 to 2003. Each line represents normalized passenger miles for a given a year. The plot shows that during the study period when public transportation usage was at its highest levels in the Tampa–St. Petersburg area, public transportation use was at its lowest levels in the Virginia Beach region. Parallel plots can be used to show the relationships among similar variables, as well as differing trends across the regions.

Below is a step-by-step explanation of how to construct parallel plots:

1. For each variable chosen, make a vertical line (in our example the variables are the four locations). All lines should be of the same length, have the same center point, and be parallel. See Figure 2.8. (JMP does this automatically.)

2. Scale each variable so that the smallest value is 0 and the largest value is 1.

3. Place a dot on the parallel line indicating the level for each variable.

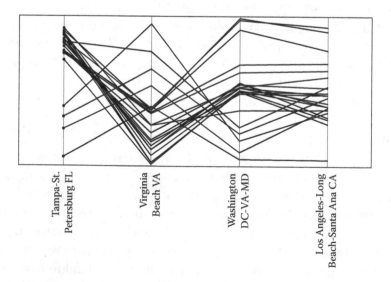

FIGURE 2.8   Parallel plot of the passenger miles traveled in each of four metropolitan areas from 1982 to 2003.

4. For each observation (row of data), connect the dots among adjacent variables with a line segment. See Figure 2.8.

From parallel plots, we can see how individual values vary across axes. In our case, each connected line represents a reading across years and shows the values for each of four urban locations. While a scatter plot matrix can show the overall relationship, or linearity, among responses, the connection among points across cells is lost. Parallel plots have the advantage of showing relationships, such as positive or negative correlation, across many variables in a comprehensible way.

2.7 TIME SERIES PLOT

The time series plot is a useful tool for displaying temporal data. It shows how variables change over time (t_i) and is a scatter plot ($t_i, y(t_i)$) with adjacent points connected by a line segment. For example, Figure 2.9 shows how passenger miles have approximately tripled in Austin, Texas, over a little more than two decades.

In this example there is a fairly linear relationship between Vehicle Miles Traveled (VMT) and time. However, it would be erroneous to assume that this relationship will continue in the future. At some point capacity issues occur and the system cannot easily handle extra travel. At this point the rate of increase, while still increasing with population, will in all likelihood be at a decreasing rate. Accordingly, while graphing does provide some insight, it is always dangerous to blindly extrapolate beyond the limits of the data.

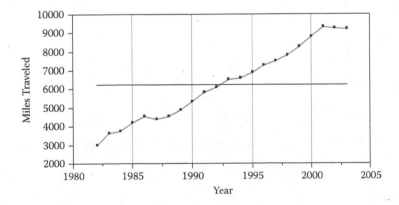

FIGURE 2.9   Time series plot of passenger miles traveled in Austin, Texas, over years.

2.8 QUALITY CONTROL PLOTS

Quality control plots are the final topic of this chapter. Quality control is a field in its own right. An excellent reference for detailed information in this area is Montgomery (2004).* One of the basic quality control plots is the Xbar chart, which is used to spot unusual data (if there are any) and is of importance to those monitoring the quality of a process. Figure 2.10 shows the Xbar chart for eight probe-vehicle drivers used in a floating car study in the Houston area. These types of studies are designed to obtain speeds in average conditions on the highway. It is important to ensure that the speeds measured from the floating cars are appropriate estimates of the average speed on the network. The plot in Figure 2.10 shows that none of the probe-vehicle drivers had unusual travel times, as is evidenced by the fact that none of the drivers had a deviation from the overall average of more than ± 3 (sample) standard deviations (calculated separately for each driver) from the sample mean. Any point outside of these limits would be unexpected. Unexpected points require more study and explanation when they occur.

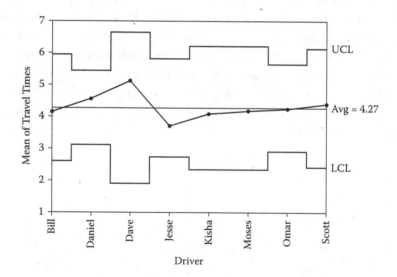

FIGURE 2.10 Xbar quality control chart of average speeds for each of eight drivers in Houston, Texas.

---

* These plots may be commonly used for quality control in pavement monitoring and measurements, and thus while not used as frequently as other data display, such plots are still relevant to transportation.

There are many different ways to construct control charts. The method given below should be thought of as an example. JMP uses a different default method for calculation of the sigma used to construct upper and lower limits. For the control chart above, separate estimates of standard deviation are used for each driver. The method below allows a single-pooled variance estimate to be used. There are many variants for control charting, and transportation engineers are encouraged to use the most meaningful method for their data. Refer to the JMP manual for more detail.

Steps for constructing an Xbar S control chart assuming equal number of replicates and a uniform process for each time period ($n_{t_i} = n, i = 1,\ldots,k$) are as follows:

1. For each time period or data segment of interest calculate the sample mean and sample variance $\bar{y}_{t_i} = \sum_{j=1}^{n} y_j / n$ and $S_{t_i}^2 = \sum_{j=1}^{n}(y_j - \bar{y}_{t_i})^2 / (n-1)$. Note that it is traditional in quality control circles to calculate a different estimator for the standard deviation. Readers are referred to the JMP manual to see how it is typically done.

2. Calculate the grand mean and pooled variance: $\bar{y} = \sum_i n\bar{y}_{t_i} / kn$ and $S_p^2 = \sum_i S_{t_i}^2 / k$, respectively. If the data segments are from different probability distributions, for example, from different drivers, then the variance estimates should not be pooled. In the latter case, individual estimates of variance would be used in the corresponding segments.

3. The upper control limit (UCL) is $\bar{y} + 3\sqrt{S_p^2} / \sqrt{n}$ and the lower control limit (LCL) is $\bar{y} - 3\sqrt{S_p^2} / \sqrt{n}$.

4. Plot the overall mean and the upper and lower control limits as shown in Figure 2.10.

5. Plot the sample averages $\bar{y}_{t_i}$ on the graph; points beyond the control limits are said to indicate an out-of-control process.

In the past, these types of quality control charts have tended not to be taught to transportation engineers since they were interpreted as more useful for industrial applications, such as manufacturing goods. However, as state and federal DOTs have begun outsourcing much of their operations, maintenance, and construction duties to outsider contractors, these types of plots are becoming increasingly common in gauging how well contractors are performing their duties.

2.9 CONCLUDING REMARKS

As previously stated, it is the authors' belief that the first step in any statistical analysis for transportation applications is to graph the data. Unfortunately, it is not always clear what form should be used first, and for this reason, it may be necessary to try different graphing types. The main goal, of course, is to develop a greater understanding of the data through the use of graphs. This comprehension might be of the relationships between the various variables (e.g., speed and volume), the nature of the data (type of pdf), whether the data include outliers (e.g., high-speed vehicles), or even whether the data are accurate (e.g., negative speeds). While this may be done from looking at the raw transportation data, it is almost impossible to do this in many applications. For example, there are over one thousand inductance loops in San Antonio, Texas, collecting data at continuous 30-second intervals. Over one year there would be approximately 1 million data points per pair of inductance loops. In this scenario, graphical data analysis would help identify the trends in the data much easier than a table would.

With the advent of readily accessible statistical software tools, it is very easy to conduct sophisticated statistical tests (as we will illustrate in the coming chapters). However, without a good knowledge of the data obtained through graphical analysis, it is very easy, particularly for novice analysts, to make fundamental mistakes. It is always important to have a good understanding of the data that you are analyzing before any statistical techniques are tried. It is our contention that graphing the data can lead to a more comprehensive understanding of them and to better statistical analyses.

HOMEWORK PROBLEMS

1. Using the headway data, plot histograms with ten, fifteen, and twenty groups or bins.

2. Using the headway data, plot a histogram of the headway data and overlay a fitted normal curve.

3. For the mobility analysis data, using only the data from very large urban areas, construct a box plot using JMP. Construct a box plot for annual hours of delay with the confidence diamond for the mean turned off. (Right clicking on the confidence diamond does this.)

4. For column 1 of the speed data, in the file "Houston speed data," construct a normal q-q plot or normal probability plot of the data using JMP. Using the confidence intervals, what can you say about the normality of the underlying probability distribution that generated the data?

5. Using the file "Beaumontmobility," construct a scatter plot of the freeway miles (on the x-axis) and total system miles (on the y-axis) using JMP.

6. Add a fitted straight line to the scatter plot constructed in question 5. Add a smoothing spline to this plot. What information do you derive from the plot?

7. Using the file "Beaumontmobility," construct a scatter plot matrix of the freeway miles and total system miles, annual hours of delay, and population density using JMP. What message do you get from the plots?

8. Using the urban mobility data file, construct a parallel plot for the travel time index for five locations of your choice. What conclusions do you arrive at?

9. Using the urban mobility data file, construct four time series plot for the travel time index for four locations of your choice. What do you think this indicates?

10. What do plots provide that a table of data does not?

REFERENCES

Leinhardt, G., and S. Leinhardt. 1980. Exploratory data analysis: New tools for the analysis of empirical data. *Review of Research in Education* 8:85–157.

Montgomery, D. C. 2004. *Introduction to statistical quality control.* 5th ed. New York: John Wiley & Sons.

SAS Institute. 2009. *JMP.* Version 8. Cary, NC: SAS Institute.

Theus, M., and S. Urbanek. 2008. *Interactive graphics for data analysis: Principles and examples.* Boca Raton, FL: CRC Press.

Tukey, J. W. 1977. *Exploratory data analysis.* Reading, MA: Addison-Wesley.

CHAPTER 3

Numerical Summary Measures

3.1 INTRODUCTION

As transportation professionals, we are often asked specific questions on the construction, operations, and maintenance of the transportation system. For example:

- Was the pavement constructed to the proper specifications?

- Are cars going too fast on a highway?

- What is the average travel time from point A to point B?

- What is the capacity of a canal lock?

In many cases, we can derive analytical solutions to these questions based on first principles. Often, however, the answer has to be derived from empirical measurement. In this situation, the engineer will need to collect data in the field to help answer the question. After plotting, the first step is to quantify the data. In this chapter, we present several numerical summary measures for summarizing the available data in a useful and informative manner. This will include measures describing the center of a data set, other locations in a data set, variability in a data set, and the relationship between two variables in a bivariate data set. The appropriate summary measures vary with the type of data, which are either categorical or numerical. Furthermore, if numerical, the data can be continuous or discrete.

3.2 MEASURES OF CENTRAL TENDENCY

As the name implies, these are measures related to the center location of the data, which can be viewed as a representative of the values in the set. The most popular measures are the mean, median, and mode, as discussed in the following sections.

3.2.1 Mean

The most popular measure of center (or central tendency) is the mean. The mean is the simple arithmetic average of the values for a numerical variable. The sample mean ($\bar{x}$) can be defined as follows:

$$\bar{x} = \frac{\sum_{i=1}^{n} x_i}{n}$$

where n is the number of observations in the data set and x_i is a value of the variable. The variable might be the speeds of vehicles, the number of traffic accidents, the lengths of pavement cracking per road segment, etc. The mean serves as a useful measure of the center. This is especially true when the data are symmetrical and unimodal. Although the mean can still be computed for bimodal or multimodal data, the meaning of the center for those data becomes unclear and, accordingly, so does the role of the mean. Of course, the exploration of the data through plots should precede the use of the numerical summary statistics. Using numerical summary measures blindly—in other words, without knowing the overall shape of the data—may lead to uninformative or even misleading summaries.

Figure 3.1 contains a frequency histogram of 204 speed measurements obtained from a TransGuide inductance loop detector located in San Antonio (SA).

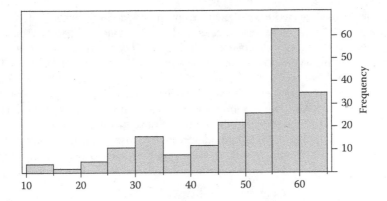

FIGURE 3.1   Frequency histogram of SA loop speed data.

Example 3.1

The sample mean ($\bar{x}$) of the 204 speed measurements shown in the histogram of Figure 3.1 is given by the sum of speed measurements in the sample divided by the number of speed measurements in the sample, which can be computed as follows:

$$\bar{x} = \frac{\sum_{i=1}^{n} x_i}{n} = \frac{10,008}{204} = 49.059.$$

Remark 3.1

In transportation lexicon the speed calculated in Example 3.1 is known as the time mean speed (TMS) because it is the calculated average over time at a single location. Because the observations are easily obtained by a single observer (or more likely a single detector), TMS is often used in practice. Another commonly used average is the harmonic mean, and the equation (when calculating the harmonic means speed) is given below:

$$SMS = S_S = \frac{d}{\frac{1}{n}\sum_{i=1}^{n} t_i} = \frac{1}{\frac{1}{n}\sum_{i=1}^{n} \frac{1}{d} t_i} = \frac{1}{\frac{1}{n}\sum_{i=1}^{n} \frac{1}{S_i}}$$

where d is the distance over which travel time is measured for each vehicle i (mi), t_i is the travel time over distance d for vehicle i (h), S_i is the speed for vehicle i to traverse distance d (mph), and n is the number of vehicles that are observed.

This is also referred to as the space mean speed (SMS) because it is the average speed of vehicles over a given distance (or space). When transportation engineers use average speed, such as in the *Highway Capacity Manual*, it is usually this SMS or harmonic average that they are referring to. The SMS is a weighted average in which each vehicle's speed is weighted by the time that it spends in the defined roadway segment or space. For example, a slower vehicle will receive more weight in the SMS calculation than a faster vehicle because it spends more time in the section. Consequently, the TMS is always greater than or equal to the SMS. Additionally, because one needs to know the time spent in each section, the SMS takes more resources to measure (e.g., two observers or detectors that can identify and track specific vehicles).

3.2.2 Median

The median is another commonly used measure of the center that is defined as the middle value in the ordered data. It is the value such that 50% of the observations in the data set are less than or equal to that value. The procedure for computing the sample median ($\tilde{x}$) is as follows:

1. Order the observations in the data from smallest to largest with any repeated values included.

2. If the number of observations in the data is odd, then the sample median is the middle entry in the ordered list. If the number of observations in the data is even, then the sample median is the average of the two middle values in the ordered list.

The median is favored over the mean when there are outliers because it is less sensitive to extremes in data. In contrast, the mean, by its very definition, is sensitive to even a single outlier because each observation is weighted equally.

Example 3.2

Because the number of observations in the data presented in Figure 3.1 is even (204), the sample median is given as the average of the 102nd and 103rd observations in the sorted values:

$$\tilde{x} = \frac{54.663 + 54.809}{2} = 54.736.$$

Note that the sample median is significantly larger than the sample mean for these data because the distribution is skewed left (there are outliers in the left tail) and the value of the mean is biased toward those outliers. Because the distribution is skewed, the sample median is a better estimate of central tendency for these data.

3.2.3 Trimmed Mean

The trimmed mean is a compromise between the mean and median. It is defined as the mean of the remaining data after a specified proportion of the smallest and the largest observations are removed from the data. The procedure for computing the 5% trimmed mean ($\bar{x}_{5\%}$), for example, is as follows:

1. Order the data values from smallest to largest.

2. Delete 5% of values from each end of the ordered list.

3. Average the remaining 90% of the values.

The trimmed mean is less sensitive to outliers than the mean, because it is inclusive of all data expect for the most extreme values. Moreover, the trimmed mean greatly contrasts with the median, which ignores all other data in favor of the middle value (or, in an even configuration, two values).

Example 3.3

The 10% trimmed mean ($\bar{x}_{10\%}$) for the data in Figure 3.1 is given by deleting 20 ($204 \times 0.1 = 20.4 \approx 20$) observations from each end of the ordered speed measurements and then taking the average of the remaining 164 values:

$$\bar{x}_{10\%} = \frac{8326.149}{164} = 50.769.$$

Note that the 10% trimmed mean lies between the sample mean and the sample median for these data. If the distribution is roughly symmetric and there is no outlier, then the mean, the trimmed means, and the median all will be close to one another.

3.2.4 Mode

The mode is the value that occurs the most frequently in the data. Unlike the mean and the median, which can be computed for numerical data only, the mode can be obtained for both numerical and categorical data. In the most rigorous interpretation of the term, only one mode can be defined within a given data set. If two values occur most often with the same frequency, then, in the strictest sense, there is no mode. However, when describing the shape of the data distribution, a mode is sometimes interpreted as a bump, and thereby in this interpretation more than one mode can be defined. The terms *unimodal*, *bimodal*, and *multimodal* are used based on such an explanation and are common to the transportation lexicon. For example, the travel time distribution of vehicles passing through a signalized intersection is often referred to as bimodal because there are two distinct bumps: those representing vehicles that are stopped by the traffic signal and those that are not.

Example 3.4

For the data in Figure 3.1, the sample mode appears to be between 55 and 60.

3.2.5 Proportion

Proportions can be obtained for the various categories of the categorical data. The sample proportions are computed as the number of responses that are in the corresponding categories divided by the total number of observations in the data. When there are only two possible categories, the categories can be named "success" and "failure" and the responses in the data can be labeled as 1 for success and 0 for failure. The sample proportion (p) of such data, which is the number of 1s divided by the number of observations in the sample, can also be viewed as the sample average of responses labeled as 1s and 0s.

Example 3.5

In addition to the traffic data, the San Antonio TransGuide system also collects incident data. In this data set, information regarding the type of incident is classified into one of four categories: major accident, minor accident, stalled vehicle, and debris. According to the TransGuide system, there were 866 incidents in May 2004 (Quiroga et al., 2004). Among those 866 incidents, 287 correspond to major accidents, 221 to minor accidents, 295 to stalled vehicles, and 63 to debris. The proportion of each incident type can be computed as

$$p_{Major} = \frac{287}{866} = 0.3314$$

$$p_{Minor} = \frac{221}{866} = 0.2552$$

$$p_{Stalled} = \frac{295}{866} = 0.3406$$

$$p_{Debris} = \frac{63}{866} = 0.0727.$$

Thus, 33% of the incidents during May 2004 were major accidents, 26% minor accidents, 34% stalled vehicles, and 7% debris.

3.3 MEASURES OF RELATIVE STANDING

Sometimes we may be interested in other (relative) locations in the data, not just the center location. Percentiles and quartiles are often used to describe other locations of interest in the data.

3.3.1 Percentile

Percentiles are regarded as the measures of relative standing because they describe relative locations compared to other observations in the data. The (100p)th percentile for any number p between 0 and 1 is defined as the value that 100 × p percent of the observations in the data set lay at or below that value. For example, the 85th percentile of the speed data in Figure 3.1 is the value that 85% of the speed measurements are below or at that value. Percentiles are computable using JMP or other statistics packages. For the sample of size n, the ith ($i = 1,...,n$) order statistic is the 100[i/(n + 1)]th percentile.

In transportation engineering, the 85th percentile speed is often used for design purposes. For example, according to the *Manual on Uniform Traffic Control Devices* (MUTCD) speed limits should be set as close to possible to the 85th percentile speed. The primary concern with setting the speed limit lower than this value is that drivers may be encouraged to not comply with this restriction by authorities. In the short run, this will lead to unsafe driving conditions, and in the long run, it will lead to a general disregard for all speed limit signs. It should be noted that there was no physical reason the 85th percentile was chosen; rather, it was assumed that most drivers are reasonable and prudent, and as such, they will choose to drive at a safe speed. Moreover, it is important to note that the 85th percentile speed is also used in sight design calculations as described in the *AASHTO Green Book: A Policy on Geometric Design of Highways and Streets*. Another common example is that roadways are often designed not for the worst demand, but rather for some percentile. This is very common in transportation systems that, by definition, are "fail soft." That is, a failure (demand greater than supply) does not generally lead to catastrophic conditions. This can be contrasted to "fail hard" systems such as bridges. For example the 100th highest hourly volume (out of 8,760 hours per year) is often used for identifying the demand for a particular roadway. It would be prohibitively expensive to design for the highest demand hourly volume, and the hundredth is seen as a compromise.

Example 3.6

For the data set of Figure 3.1, the 85th percentile is given as 60.401.

3.3.2 Quartile

Quartiles are the particular percentiles. They divide the data into quarters; therefore, the percentiles are the 25th, the 50th, and the 75th. The 25th percentile and the 75th percentile are usually referred to as the lower quartile and the upper, respectively. The 50th percentile, which is the middle quartile, is the same as the median.

Example 3.7

For the data set of Figure 3.1, the 25th percentile and the 75th percentile are 41.838 and 58.533, respectively.

3.4 MEASURES OF VARIABILITY

So far, we have discussed the various measures of locations in the data. Another important feature of the distribution of the data is variability, or the extent of the spread of values in the data. Metrics that attempt to measure variability are sometimes referred to as measures of dispersion.

3.4.1 Range

The simplest measure of variability is the range. It is defined as the difference between the maximum and minimum values in the data. Because the range is determined by the two most extreme values in the data, it is extremely sensitive to even a single outlier and generally increases as sample size increases. Thus, the use of range as a measure of variability should be avoided when there is an outlier or when comparing variability of samples that have different numbers of observations.

Example 3.8

For the data set of Figure 3.1, the sample range is computed as

Range = maximum − minimum = 63.767 − 13.304 = 50.4630.

3.4.2 Variance

In describing variability in the data, it is also important to consider how the other values, not just the two endpoints, in the data are distributed. The variance, which is the average of the sum of squared deviations of the observations from the mean, uses all the observations. The sample variance (s^2) is defined as

$$s^2 = \frac{\sum_{i=1}^{n}(x_i - \bar{x})^2}{n-1}$$

where n is the number of observations in the data set and $\bar{x}$ is the sample mean. It may be seen that the farther a particular observation is from the mean, the larger the value of s^2 will be. The variance, however, is not a measure of spread because its units are the square of the data units.

Example 3.9

For the data set of Figure 3.1, the sample variance can be computed as

$$s^2 = \frac{\sum_{i=1}^{n}(x_i - \bar{x})^2}{n-1} = \frac{32{,}125.310}{203} = 158.253.$$

3.4.3 Standard Deviation

The standard deviation—defined as the square root of the variance—is a widely used measure of variability and has the same units as the original data, such as miles per hour or meters. It can be interpreted as a measure of a typical deviation of an observation from the mean.

Example 3.10

For the data set of Figure 3.1, the sample standard deviation can be computed as

$$s = \sqrt{s^2} = \sqrt{158.253} = 12.580.$$

One or two unusually small or large observations can easily inflate the values of s^2 and s even though there may, in fact, be little variability in the rest of the observations. Although they are not as sensitive as the range, both the variance and the standard deviation are sensitive to outliers and are affected by extremes in data.

3.4.4 Interquartile Range

Less susceptible to the influence of outliers, the interquartile range (IQR) is another measure of variability. The interquartile range is defined as the difference between the upper quartile and the lower quartile. The outliers cannot affect the value of the IQR because it is the range of the middle 50% of the data.

When the data maintain approximately a normal distribution, there exists the following relationship between the standard deviation and the IQR:

$$s = \frac{IQR}{1.35} \approx \frac{3}{4} IQR.$$

If s is much larger than $IQR/1.35$, then this indicates that the distribution may give more weight to observations far from the center in comparison to the normal distribution. This would indicate that there may be outliers in the data.

Example 3.11

For the data set of Figure 3.1, the sample interquartile range can be computed as

IQR = upper quartile – lower quartile = 58.533 – 41.838 = 16.695.

3.4.5 Coefficient of Variation

Another measure of variability that is often employed in transportation engineering is the coefficient of variation (CV). The coefficient of variation expresses the standard deviation as a percent of the mean as follows:

$$CV = \frac{100s}{\bar{x}}.$$

The CV is a measure of relative variation of the data when the measure of dispersion is relative to the measure of central tendency. One common interpretation of the CV is that the values in the data set lie, on average, within approximately *CV* percent of the mean. The coefficient of variation is especially useful when comparing the variability of the data sets measured by different units because it is unit-free, and so it automatically takes into account the different units in different data sets. It can also be estimated as the standard deviation of the log of the data.

The CV is often used in engineering because it combines both a measure of central tendency and a measure of dispersion into a single metric. Often it is hard to interpret one metric without knowing the value of another. For example, a standard deviation of ten minutes has a different meaning on a ten-minute trip vs. a ten-hour trip. In the former, a typical driver would consider the trip to have considerable variability, while in the latter almost none. In addition, in some transportation applications it is often found that the relationship between the mean and standard deviation is approximately constant over time and space. Consequently, if a transportation professional only has one measure, the other can be readily estimated if the CV from a previous study is available.

Example 3.12

For the data set of Figure 3.1, the sample coefficient of variation can be computed as

$$CV = \frac{100s}{\bar{x}} = \frac{100 \times 12.580}{49.060} = 25.642.$$

3.5 MEASURES OF ASSOCIATION

So far we have discussed various numerical measures for summarizing univariate data. However, transportation professionals often consider more than one variable in the data set, and the relationship among those variables is of interest. The correlation coefficient (or Pearson correlation coefficient) is a measure of the strength and direction of the relationship between two numerical variables, and is often used to describe and assess the strength of a relationship. There are two widely used correlation coefficients: Pearson's correlation coefficient and Spearman's rank correlation coefficient.

3.5.1 Pearson's Sample Correlation Coefficient

Let (x_1, y_1), (x_2, y_2), ..., (x_n, y_n) denote a sample of measurements on the variables X and Y, where $\bar{x}$ and $\bar{y}$ are the corresponding sample means. For example, let X represent spot speed on a roadway and Y the volume. Pearson's sample correlation coefficient is defined as

$$r = \frac{\sum_{i=1}^{n}(x_i - \bar{x})(y_i - \bar{y})}{\sqrt{\sum_{i=1}^{n}(x_i - \bar{x})^2}\sqrt{\sum_{i=1}^{n}(y_i - \bar{y})^2}}$$

and it is a measure of linear relationship between two variables X and Y. Note that the value of r is between –1 and 1, and it is unit-free. A positive value of r indicates an increasing relationship between variables, and a negative value indicates a decreasing relationship between variables. A value of r close to –1 indicates a strong negative linear relationship, and a value close to 1 indicates a strong positive linear relationship. Because it only measures the extent of the linear relationship, a value of r close to 0 does not rule out any strong nonlinear relationship between X and Y. It could be very misleading to conclude that there is no relationship based on a value of r close to 0 without first examining a scatter plot of the bivariate data.

Example 3.13

Table 3.1 contains the volumes measured at fourteen locations along a corridor at two different time periods. Figure 3.2 shows the scatter plot of the data.
    For the data set given above, Pearson's sample correlation coefficient can be computed as

$$r = \frac{\sum_{i=1}^{n}(x_i - \bar{x})(y_i - \bar{y})}{\sqrt{\sum_{i=1}^{n}(x_i - \bar{x})^2}\sqrt{\sum_{i=1}^{n}(y_i - \bar{y})^2}} = \frac{104{,}337}{303.0319 \times 348.5547} = 0.9878.$$

Given the scatter plot, it is not surprising that Pearson's correlation coefficient statistic indicates a very strong linear relationship in the data.

TABLE 3.1   Traffic Volumes Measured at Different
Locations along a Corridor

Location	Volumes at Time 1	Volumes at Time 2
1	303	341
2	56	68
3	136	177
4	146	188
5	7	8
6	32	17
7	35	51
8	50	51
9	21	22
10	12	7
11	18	24
12	110	96
13	169	169
14	28	30

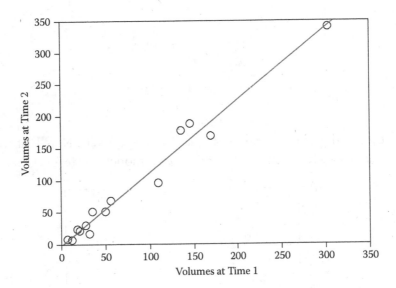

FIGURE 3.2   Scatter plot of volumes at time 1 and volumes at time 2.

3.5.2 Spearman's Rank Correlation Coefficient

Recall that Pearson's correlation coefficient is not a useful measure of association when the relationship between two variables is nonlinear. Another potential drawback to Pearson's correlation coefficient is that it is sensitive to outliers. Spearman's rank correlation coefficient is a good alternative to

Pearson's correlation coefficient when there is an outlier in the data, or the relationship between the variables is not linear, but still monotonic.

As a concept, Spearman's rank correlation coefficient is essentially the same as Pearson's correlation coefficient. The difference is that the ranks of observations are used, rather than actual observations. Thus, to calculate Spearman's correlation between paired variables— (x_1, y_1), (x_2, y_2), ..., (x_n, y_n)—the data are ranked separately, and then the Pearson's correlation coefficient is computed for the ranks. For example, suppose our data consist of pairs (2, 5), (3, 9), (5, 10), (7, 10.1), and (9, 10.2). Then the ranked data are: (1, 1), (2, 2), (3, 3), (4, 4), and (5, 5). While the Spearman's correlation coefficient equals 1, the Pearson's correlation coefficient, based on the original data, is computed to be $r = 0.756$. Figure 3.3 presents a scatter plot of the original data. The scatter plot shows that while there is a strong monotonic relationship between the two variables, it is not strictly linear; in fact, it is highly nonlinear.

The formula for the Spearman's rank correlation coefficient can also be given as

$$r_S = \frac{\displaystyle\sum_{i=1}^{n} R_{x_i} R_{y_i} - \dfrac{n(n+1)^2}{4}}{\dfrac{n(n-1)(n+1)}{12}}$$

where R_{x_i} and R_{y_i} are the ranks of x_i and y_i, respectively. Note that when the relationship between two variables is not monotonic (e.g., U shaped),

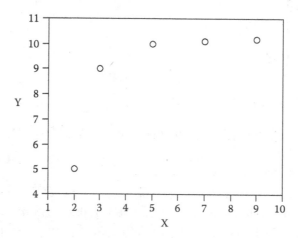

FIGURE 3.3 Scatter plot of the original data.

Spearman's correlation coefficient may not be an appropriate measure of association either.

Example 3.14

For the data set of Example 3.13, Spearman's sample correlation coefficient can be computed as

$$r_S = \frac{\sum\limits_{i=1}^{n} R_{x_i} R_{y_i} - \frac{n(n+1)^2}{4}}{\frac{n(n-1)(n+1)}{12}} = \frac{1004 - 787.50}{227.50} = 0.9516.$$

For this data set, the difference between Pearson's correlation coefficient and Spearman's correlation coefficient is relatively small because there are no outliers and the relationship between two variables is almost linear, as can be observed from Figure 3.2.

Remark 3.2

Although the procedures or formulas for computing summary statistics are provided to help readers understand the concepts in this book, in practice those statistics can be more easily computed by software packages, such as JMP. It is rare to do these calculations by hand. However, similar to most engineering applications, it is unwise to use the computer program without a deep understanding of the logic and theory behind the calculations.

3.6 CONCLUDING REMARKS

What is the average? How much spread is there in the data? A generation ago, this was the extent of statistical knowledge most practicing engineers would need to know. Yet, as this chapter demonstrated through explaining descriptions for data sets, today's engineer requires a broader array of descriptive statistics. With changes in the industry, the transportation engineer's job is much more complex. The modern transportation engineer now has to use the complete set of descriptive statistics to communicate effectively. Thus, the field has changed from just utilizing means and standard deviations to relying on more sophisticated descriptions in order to communicate important results. In a world where the transportation budgets are not keeping up with demands, the necessity for analyzing options, from a valid statistical perspective, has never been more important.

HOMEWORK PROBLEMS

1. Using the highway speed data of Figure 2.2, compute the sample mean (time mean speed) and the sample median. Do those values differ from each other? What would you recommend as a measure of center for these data? Why?

2. The accompanying data are the travel times (min) on an arterial street collected by ten test vehicles.

Vehicle	1	2	3	4	5	6	7	8	9	10
Travel Time (min)	10	14	9	10	12	16	10	11	22	11

a. Calculate the values of the sample mean, median, 5% trimmed mean, and mode.

b. From the relative locations of the above summary statistics, can you conjecture the shape of the distribution (smoothed histogram)? Would it be symmetric, skewed right, or skewed left?

c. What is the proportion of vehicles of which travel time exceeds twenty minutes based on the above data?

d. Compute the 25th percentile and the 75th percentile.

e. Compute the sample standard deviation and the interquartile range. What would you recommend as a measure of variability for these data? Why?

3. The accompanying data are the vehicle speeds (ft/s) and lateral positions (in.) measured for fifteen vehicles at four locations along a curve: upstream location (U), advance curve warning sign location (W), point-of-curve location (PC), and midpoint-of-curve location (MC). This data set is saved as a JMP file named "vehicle speed and lateral position" in the data CD. It was extracted for illustration purposes from a larger data set obtained from the FHWA project, "Pavement Marking Demonstration Projects: State of Alaska and State of Tennessee" (see Carlson et al., 2010).

Vehicle No.	Speed at U	Speed at W	Speed at PC	Speed at MC	Lateral Position at U	Lateral Position at W	Lateral Position at PC	Lateral Position at MC
1	62	63	62	54	53	28	13	53
2	70	64	62	50	33	41	17	42
3	70	67	64	61	44	48	23	54
4	70	71	68	46	43	24	13	43
5	73	75	70	65	29	40	19	62
6	59	65	62	58	35	28	25	51
7	56	56	46	47	32	48	14	36
8	59	60	57	51	34	37	14	33
9	61	64	62	50	28	33	15	71
10	61	62	61	52	23	32	14	36
11	65	66	63	56	38	26	15	56
12	57	81	70	64	43	41	19	59
13	46	71	67	60	50	46	25	52
14	68	66	62	56	43	28	11	51
15	59	51	53	48	49	58	14	77

a. Calculate the values of the sample mean and standard deviations for speed at MC and lateral position at MC and comment on them. How would you compare the variability for speed at MC to that for lateral position at MC? Is comparing the two standard deviations meaningful? Explain.

b. Calculate the values of CV for speed at MC and lateral position at MC, respectively. How do the values compare? Interpret the results.

c. Calculate the value of the Pearson's correlation coefficient between speed at W and speed at PC. How would you describe the relationship between two variables?

d. Calculate the value of the Spearman's correlation coefficient between speed at W and speed at PC. Is the value larger or smaller than the Pearson's correlation coefficient? If the two values are not very close, what do you think causes the main difference?

e. Construct a scatter plot of speed at W and speed at PC. Identify any potential outlier in this bivariate data. Remove any outliers from the data and compute the Pearson's correlation coefficient and the Spearman's correlation coefficient again. What do you observe?

REFERENCES

Carlson, P. J. et al. 2010. *Pavement marking demonstration projects: State of Alaska and State of Tennessee.* FHWA-HRT-09-039. Washington, D.C. Federal Highway Administration.

Quiroga, C., E. Kraus, R. Pina, K. Hamad, and E. S. Park. 2004. *Incident characteristics and impact on freeway traffic.* Research Report 0-4745-1. Austin, TX: Texas Department of Transportation.

CHAPTER 4

Probability and Random Variables

4.1 INTRODUCTION

Transportation engineering is concerned with planning, designing, building, operating, and maintaining all aspects of the transportation system. To accomplish this, transportation engineers must have a deep understanding of the demands on the system (number of vehicles wanting to use a highway, number of trains on railway tracks, number of ships in a port), the supply of the system (number of roads, miles of railway tracks, traffic signal settings), and their interactions. This is a very complicated process that is compounded by the random or probabilistic nature of the variables. As engineers we tend to handle this complexity and randomness by creating simplified models of the phenomena of interest.

Probability theory is the calculus for chance occurrences. It provides language and grammar for computing the arithmetic of chance. Careful application of the rules will allow the reader to obtain the correct answer. This section will explain the use of basic probability concepts that are used in transportation engineering applications.

4.2 SAMPLE SPACES AND EVENTS

Statisticians often use the following definitions to develop the terminology of probability. A sample space contains all the possible outcomes for which probabilities are needed. For example, the sample space for roadway vehicles in the United States is the approximately 150 million licensed vehicles.

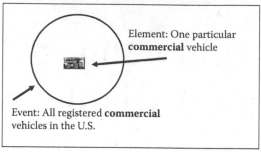

FIGURE 4.1 Sample space, event, and element for commercial vehicles in the United States.

The simplest events, which essentially are building blocks of things for which we will want to compute probabilities, are called elements. For example, a commercial vehicle. The set of 90+ million commercial vehicles in the U.S. would be an event. The collection of elements that we want to compute the probability for is called an event. These definitions are demonstrated in Figure 4.1.

4.3 INTERPRETATION OF PROBABILITY

Let A and B be two events in the sample space, Ω. For example, the sample space may be all vehicles traversing a roadway in a given time interval t. Event A might be the observation of a tractor-trailer combination and event B might be the observation of a motorcycle. We are going to define a general probability measure or function for each event. The probability function assigns a probability, P, or number, to every event. The rules, which mathematicians refer to as axioms, are as follows:

1. For any event, A is $0 \le P(A) \le 1$.

2. $P(\Omega) = 1$, and $P(\emptyset) = 0$, where $\emptyset$ denotes the empty event (set).

3. If A and B are disjoint, that is, $A \cap B = \emptyset$, then $P(A \cup B) = P(A) + P(B)$. More generally if $A_1, A_2, \ldots$ is a sequence of mutually disjoint events, then $P(A_1 \cup A_2 \ldots) = P(A_1) + P(A_2) + \ldots$.

It can be shown that as a consequence of rules 1 to 3, rules 4 and 5 follow:

4. $P(A^c) = P(\Omega) - P(A) = 1 - P(A)$, where A^c is the complementary event to A in Ω.
5. $P(A \cup B) = P(A) + P(B) - P(A \cap B)$.

These rules, while simple, are the basis of all the probability calculations that will be done in this textbook. Additional references that students may wish to use are Ott and Longnecker (2009) and Devore (2008). Examples 4.1 and 4.2 are simple scenarios to familiarize yourselves with the rules.

Example 4.1

What is the chance that a die is thrown and the number that comes up is greater than 4?

The answer can be obtained using the rules or by intuition. The sample events are 1, 2, 3, 4, 5, and 6. The numbers greater than 4 are 5 and 6. If it is a fair die, each number has a probability of 1/6. Using axiom 3, the answer is 1/6 + 1/6 = 1/3.

Example 4.2

Now, suppose that we roll a pair of dice and we want the probability that the total is an even number, but we also require that neither die contains an odd number.

We can see from Table 4.1 that there are thirty-six possible outcomes and each is equally likely.

It can be seen that the entries corresponding to an even column number and even row number contain the events of interest: those events that sum to an even number, but contain no odd number from any die. It is also clear that of the thirty-six equally likely outcomes, there are nine that satisfy the criterion. Consequently, the probability of this event is 9/36, or 1/4.

TABLE 4.1   Possible Outcomes from a Pair of Fair Dice

(1, 1)	(1, 2)	(1, 3)	(1, 4)	(1, 5)	(1, 6)
(2, 1)	(2, 2)	(2, 3)	(2, 4)	(2, 5)	(2, 6)
(3, 1)	(3, 2)	(3, 3)	(3, 4)	(3, 5)	(3, 6)
(4, 1)	(4, 2)	(4, 3)	(4, 4)	(4, 5)	(4, 6)
(5, 1)	(5, 2)	(5, 3)	(5, 4)	(5, 5)	(5, 6)
(6, 1)	(6, 2)	(6, 3)	(6, 4)	(6, 5)	(6, 6)

There are two other key concepts that are yet to be defined. The first is the concept of independence. Two events A and B are independent if

$$P(A \cap B) = P(A)P(B). \tag{4.1}$$

Note that this definition could have been used in Example 4.2 because we implicitly assumed that the outcome of each die was independent. In other words, the probability that the first die was an even number was 1/2, and the probability that the second die was an even number was also 1/2. Using the rule for two independent events, the answer is the product of the two probabilities, or 1/4.

4.4 RANDOM VARIABLE

Another important definition is that of a random variable. A random variable is a function that assigns numbers to elements of the sample space. In the two previous examples, we talked about numbers on the dice, and of course, dots on a dice demarcate the corresponding value of the die. The random variable assigned the number 1 to the outcome where one dot appeared, and so on. The importance of random variables is that we need only consider probabilities of events characterized as numbers. For example, you might assign the event of a crash incident as the number 1, and in the event a vehicle does not crash, the number 0. Assume that you are examining n trips and you have an event X_i (e.g., the random variable representing the crash event) for the ith trip. If we wanted to calculate the total number of crashes in n trips, we would simply find the sum of the events:

$$\sum_{i=1}^{n} X_i. \tag{4.2}$$

Because the event is a random variable, so too is the sum. In a sense, this is a fairly obvious statement. For example, if we know the number of accidents in any given month is a random variable, then it follows that the yearly number of accidents (e.g., the sum of accidents over every month) is also a random variable. A more interesting question is whether we can derive the properties of the accidents per year if we know the properties of accidents per month. This question will be answered later in this chapter.

One way statisticians describe the probabilities of events, such as $(X \le x_0)$, where x_0 is a constant, is through the use of a cumulative distribution function (cdf). The cdf for a random variable, X, is a function of sets of the form $(X \le x_0)$ and is defined by

$$F(x_0) = P(X \le x_0). \tag{4.3}$$

The probability of $(x_1 < X \le x_2)$ is calculated by $F(x_2) - F(x_1)$. In essence, the cdf completely describes the probability that an event will occur.

Traditionally, engineering texts consider two varieties of random variables. The first are those that may take all possible values on some interval, and are referred to as continuous random variables. For example, the speed of a vehicle can take on any real value (between 0 and the vehicle's maximum speed) and may therefore be considered continuous. As we will demonstrate, there are many standard distributions that are used to describe random variables. As you progress through this text, you will see that the most common distribution is known as the normal, or Gaussian, distribution. The other type of random variables have discrete outcomes. These random variables take values from a list. For example, assume an engineer was interested in the number of vehicles arriving at an intersection. It is clear that this would be a nonnegative, integer number because there are no fractional vehicles. This example is often modeled using the Poisson distribution. Another example of a discrete random variable is the binomial distribution, which has been used to model the number of crashes in n trips (where n is a predetermined number, such as 1,000). The most common distributions used in transportation engineering are described in Chapter 5.

It cannot be emphasized enough that probability distributions models are approximations of reality. For example, you may have a measuring device that only measures speed to tenth of a mile per hour. While the events are discrete and can be listed as discrete values, speeds are typically modeled as continuous random variables. Similarly, it is clear that traffic volume measurements are discrete. However, most transportation literature treats volume as continuous. The rationale is that the approximation does not appreciably change the answer and that, as you will see in Chapter 5, continuous distributions are often much easier to use.

This brings us to the densities of a continuous random variable. The probability density function (pdf) for a random variable is the derivative of the distribution function:

$$f(x) = \frac{dF(x)}{dx} \ge 0. \tag{4.4}$$

As the name implies, the pdf (probability *density* function), unlike the cdf (cumulative *distribution* function), cannot be used to calculate probabilities of events directly. However, the area under the pdf is the probability. For example, for any set of events, or A :

$$P(X \in A) = \int_A f(x)dx. \qquad (4.5)$$

In other words, once you know the pdf, the probability of any event can be calculated by integration. Of course, this would be rather cumbersome to do by hand every time one wanted to calculate probabilities, and particularly for some of the more complicated pdfs. Most software packages will allow you to calculate this directly. Prior to computers these probabilities were precalculated and put into tables. Indeed, these tables are still taught today, but from a practical point of view, they are obsolete. Regardless, in order to calculate the probability of an event for a continuous variable, the transportation professional will have to perform an integration by using a computer program, a table of precalculated values, or by hand. Please note that the probability density function is often called the density in practice.

The discrete analog to the density is the probability mass function (pmf), or $p(x)$. In contrast to the pdf, the pmf gives the probability of a given element, or x, directly. For discrete random variables it has the property that

$$P(X \in A) = \sum_A p(x). \qquad (4.6)$$

4.4.1 Functions of a Random Variable

Transportation engineers use random variables to model many phenomena. For example, they might model speed and volume as random variables with specific properties. Often we want to calculate the associated traffic density, which is the traffic flow divided by the traffic speed. As you might expect, the function of a random variable is also a random variable. Therefore, if speed and flow are random variables, then so is traffic density. Yet, only in the most simple of cases it is possible to calculate a simple closed-form pdf or cdf of a random variable that is a function of other random variables.

To illustrate, consider X as a random variable. As discussed previously, X^2 and $Log(|X|)$ are also random variables. The distribution function for a function of a random variable X can be obtained from the distribution function of X. For example:

$$P(X^2 \leq t) = P(-\sqrt{t} \leq X \leq \sqrt{t}) = F(\sqrt{t}) - F(-\sqrt{t}) \qquad (4.7)$$

for a nonnegative scalar, t, and equals zero for negative t.

4.5 EXPECTATIONS OF RANDOM VARIABLES

The expectation of random variables is an important property for describing the location of their distribution function. Recognizing that all analogies are problematic, the concept of a body's center of gravity is an engineering-related analogy to the expectation, or mean, of a random variable. For continuous random variables the mean is defined as

$$\mu = EX = \int_{-\infty}^{\infty} xf(x)dx. \qquad (4.8)$$

For discrete random variables the expectation is defined as

$$\mu = EX = \sum_{\{x_i|p(x_i)>0\}} x_i p(x_i). \qquad (4.9)$$

It can be seen from the above formulas that, as the name implies, the expected value is the mean, or average value, of the distribution. The expected value does not necessarily have to be one of the values that could occur. It is easy to show that the expected value of a roll of a die is 3.5, which is clearly a value that would never occur on the actual die. However, over a large number of rolls, it would be expected that the average of the values of the rolls would be 3.5.

A distinction should be made that when statisticians talk about expectations they are basically discussing a weighted average of a function of the random variable where the weights are the probabilities of the given event, X. For the mean, the function is simply $f(x) = x$. However, one could take expectations of any function, and some of the more common functions will be introduced in the following pages.

Unless otherwise stated (as when we note it is nonexistent for some important random variables), we will assume that these integrals and sums exist. An example of a distribution without moments is the t-distribution with 1 degree of freedom, also known as a Cauchy distribution. In these situations, other measures, such as median and interquartile range, are used to describe the distribution rather than expectations or means.

4.5.1 Expected Values Are Linear

We will often have to compute expected values for weighted sums of random variables.

Suppose, for example, that X and Y are random variables and a and b are constants, and we want to compute the expected value of $aX + bY$. Note that the values of a and b do not matter in the context of the theory. Of course, different values will have different interpretations. If $a = b = 1/2$, then we have an average of the two random variables. On the other hand, if $a = 1$ and $b = -1$, then we have a difference. Regardless of the values of the constants, the expectation of the linear function of X and Y can be calculated as follows:

$$E(aX + bY) = aEX + bEY. \tag{4.10}$$

Transportation engineers often use formulas like the one shown above when examining functions of random variables. Often it is mathematically difficult, or even impossible, to derive the sum of two distributions. However, if the transportation engineer is really interested in the mean of the function, even though the distribution might be better, then these types of formulas are very useful.

4.6 COVARIANCES AND CORRELATION OF RANDOM VARIABLES

As discussed previously, it is often useful to calculate expectations of functions of the random variables. The expectations that are computed most often are given special names. The mean is perhaps the most commonly used expectation. However, the key point is that any expectation is a weighted value of a function where the weights are the probabilities of the event X.

The mean is used to measure the central tendency of a pdf. The variance is used to measure the spread of a pdf. For continuous distributions, the variance can be found as follows:

$$\sigma^2 = E\left(X - \mu\right)^2 = \int_{-\infty}^{\infty} \left(x - \mu\right)^2 f(x)dx. \tag{4.11}$$

For discrete random variables the variance is calculated using the formula

$$\sigma^2 = E\left(X - \mu\right)^2 = \sum_{\{x_i | p(x_i) > 0\}} \left(x_i - \mu\right)^2 p(x_i). \tag{4.12}$$

It is clear that the variance is the expectation of the square of the difference between each event, X, and the mean of the distribution. Not surprisingly, the higher the spread of the pdf, the higher the variance. There are many ways of measuring the spread of the pdf; for example, one could take the absolute values of the differences instead of the square. However, for a variety of reasons not covered in this text, the variance is the most useful and, as such, is given a special designation.

Two common measures of association for random variables X and Y are the covariance of X and Y defined as $\sigma_{XY} = E(X - EX)(Y - EY)$, and the correlation is defined as $\rho_{XY} = \sigma_{XY} / \sigma_X \sigma_Y$. The terms $\sigma_{xx} = \sigma_x^2$ and $\sigma_{yy} = \sigma_y^2$ are called the variance of X and Y, respectively.

The following properties hold for covariances and variances:

1. $Cov(aX, bY) = abCov(X, Y)$.

2. If X and Y are independent random variables, $Cov(X, Y) = 0$.

3. $Var(aX) = a^2 Var(X)$.

4. $Var(aX + bY) = a^2 Var(X) + b^2 Var(Y) + 2abCov(X, Y)$.

As before, sometimes it is impractical to calculate the pdf of a function of two random variables. In these situations the above formulas are useful if we would like to obtain measures of central tendency and spread.

4.7 COMPUTING EXPECTED VALUES OF FUNCTIONS OF RANDOM VARIABLES

Propagation of error is the name for the calculus used to obtain means and variances of nonlinear functions of random variables using Taylor expansions. The subsequent presentation follows the reference Ku (1966) closely.

Let $Z = f(X, Y)$. Then we approximate the mean of Z as

$$EZ = f(EX, EY) \approx f(\mu_x, \mu_y). \tag{4.13}$$

The variance of Z is approximated using the formula

$$\sigma_Z^2 = \left(\frac{\partial f}{\partial x}\right)^2 \sigma_X^2 + \left(\frac{\partial f}{\partial y}\right)^2 \sigma_Y^2 + 2\left(\frac{\partial f}{\partial x}\right)\left(\frac{\partial f}{\partial y}\right)\rho_{XY}\sigma_X\sigma_Y. \tag{4.14}$$

TABLE 4.2   Approximate Variances of
Transformed Random Variables Obtained by
Using Propagation of Error

Function	Approximate Variance
$\dfrac{1}{X}$	$\dfrac{\sigma_X^2}{\mu_X^4}$
$\ln X$	$\dfrac{\sigma_X^2}{\mu_X^2}$
e^X [a]	$e^{2\mu_X}\sigma_X^2$

[a] Can produce a highly skewed distribution, so the approximation may be inaccurate. This is particularly true for X having a large variance.

Using these formulas, we have the following examples: Table 4.2 shows univariate transformations of X and the resulting approximate variances.

Example 4.3

Suppose that we are interested in the space mean speed of vehicles on a 1-mile segment of Houston highway. The formula that we would use is

$$\hat{\mu}_{sms} = \frac{1}{\sum_{i=1}^{n} t_i / n} = \frac{1}{\bar{t}}. \tag{4.15}$$

In which t_i is the travel time over a specified distance measured in hours per mile, and n is the number of vehicles. Then, the variance of this estimator is approximated by $S_t^2 / \bar{t}^4$, where $\bar{t}$ is the average travel time and S_t^2 is the squared estimated standard error for $\bar{t}$.

Example 4.4

A formula that is used for calculation of density is

$$k = \frac{5280}{\bar{d}} = 5280\left(\frac{1}{\bar{d}}\right). \tag{4.16}$$

See May (1990). Here, k is density in vehicles per lane-mile, and $\bar{d}$ is average distance headway. The variance of this estimator is then approximated by $(5280)^2 S_d^2 / \bar{d}^4$, where S_d^2 is the squared estimated standard error for $\bar{d}$.

Table 4.3 shows examples of transformations of X and Y, and the resulting variances.

TABLE 4.3   Approximate Variances of Functions of Two Random Variables Obtained by Using Propagation of Error

Function	Approximate Variance If X and Y Are Independent	Term to Be Added If X and Y Are Correlated
$\dfrac{X}{Y}$	$\left(\dfrac{\mu_X}{\mu_Y}\right)^2\left(\dfrac{\sigma_X^2}{\mu_X^2}+\dfrac{\sigma_Y^2}{\mu_Y^2}\right)$	$\left(\dfrac{\mu_X}{\mu_Y}\right)^2\left(-2\dfrac{\sigma_{XY}}{\mu_X\mu_Y}\right)$
$\dfrac{X}{X+Y}$	$\left(\dfrac{1}{\mu_X+\mu_Y}\right)^4\left(\mu_Y^2\sigma_X^2+\mu_X^2\sigma_Y^2\right)$	$\left(\dfrac{1}{\mu_X+\mu_Y}\right)^4\left(-2\mu_X\mu_Y\sigma_{XY}\right)$

Example 4.5

Let ρ be traffic density, $\bar{X}$ the estimated average arrival rate, and $\bar{Y}$ the estimated mean service rate. Both means are in vehicles per time interval units. Consequently, $\rho = \bar{X}/\bar{Y}$, and assuming that the two means are independent, its estimated variance is $\left(\bar{X}/\bar{Y}\right)^2\left(S_{\bar{X}}^2/\bar{X}^2+S_{\bar{Y}}^2/\bar{Y}^2\right)$. The estimator is $S_{\bar{X}}^2 = S_X^2/n$, and is the squared estimated standard error for $\bar{X}$. Refer to May (1990) for the motivation behind the use of ρ.

4.8 CONDITIONAL PROBABILITY

Transportation professionals base decisions upon available knowledge and data. Statisticians use conditional probability to formalize the process of making decisions based upon available data. Let us look at Table 4.1 again. The probability of rolling a 6 on the roll of two dice is $5/36$. However, if we are told that the first die is a 6, then the probability of rolling a 6 is 0. For random variables X and Y, the conditional probability that X is in set A, given that Y is in set B, is defined as

$$P(X \in A \mid Y \in B) = \frac{P(X \in A, Y \in B)}{P(Y \in B)} \tag{4.17}$$

provided that $P(Y \in B) \neq 0$. In the case of the dice example,

$$P(X+Y=6 \mid Y=6) = \frac{P(X=0, Y=6)}{P(Y=6)} = 0.$$

For continuous random variables X and Y, the conditional probability density function is defined both as

$$f(x \mid y) = \frac{f_{X,Y}(x,y)}{f_Y(y)} \tag{4.18}$$

and

$$P(X \in A \mid Y = y) = \int_A f(x \mid y)dx. \tag{4.19}$$

The formulas given above provide the basic elements needed to calculate conditional probability and corresponding conditional moments. For example, the conditional mean of X given that $Y = y$ is defined as

$$E[X \mid Y = y] = \int_{-\infty}^{\infty} xf(x \mid y)dx. \tag{4.20}$$

It can be seen that if random variables X and Y are independent, then for all y:

$$f(x \mid y) = \frac{f_{X,Y}(x,y)}{f_Y(y)} = \frac{f_X(x)f_Y(y)}{f_Y(y)} = f_X(x). \tag{4.21}$$

Example 4.6

Suppose that we have twenty vehicle speeds recorded. Half of the speeds are recorded in sunny weather, and half of the speeds are recorded during a thunderstorm (Table 4.4).

From the data, we can see that the percentage of vehicles speeds traveling over 58 mph is 45% (e.g., 9/20 = .45). However, the percentage of drivers traveling over 58 mph on clear weather days is 90% (P(mph > 58 & clear) / P(clear) = 9/20/10/20 = .90. Thus, in this sample data set, the weather condition and speed are not independent variables. Many statistical procedures, such as correlation and regression, are designed to gain information by characterizing dependencies among variables.

The difference between conditional and ordinary (marginal) probabilities can be understood in the simple example based on airline travel. It can readily be shown that the chance that a given airline traveler will die in an airplane crash is low. However, conditional upon the given traveler being on an airplane that crashes, the probability of death is high.

TABLE 4.4 Vehicle Speed and Weather

Vehicle Number	Speed (mph)	Weather
1	65	Clear
2	67	Clear
3	75	Clear
4	59	Clear
5	57	Clear
6	65	Clear
7	67	Clear
8	71	Clear
9	63	Clear
10	62	Clear
11	30	Thunderstorm
12	26	Thunderstorm
13	36	Thunderstorm
14	46	Thunderstorm
15	21	Thunderstorm
16	50	Thunderstorm
17	36	Thunderstorm
18	43	Thunderstorm
19	30	Thunderstorm
20	41	Thunderstorm

4.9 BAYES' THEOREM

We may want to know the probability that a person was driving under the influence of alcohol if he or she had a positive Breathalyzer™ test. A problem with this example is that a direct calculation of this probability would be difficult. We would have to collect a random sample of drivers who had positive Breathalyzer tests and give them an immediate blood test. If they did not consent, the probabilities could not be calculated. Bayes' theorem gives us an easier and more practical way to do this type of calculation. The two main uses for Bayes' theorem are to calculate conditional probabilities that would be difficult to do correctly, and to provide improved parameter estimates that combine various sources of information. The latter use is covered in Chapter 14.

Suppose that we have a collection of events, $\{A_i, i=1,\ldots,k\}$, that are mutually exclusive and partition the sample space, such as $A_1 = $ DWI and

A_2 = not DWI. Suppose that we have a second set of mutually exclusive events that are observable, $\{B_i, i = 1, \ldots, m\}$, such as B_1 = failed Breathalyzer test and B_2 = passed Breathalyzer test. Then we have the following:

$$P(A_i \mid B_j) = \frac{P(B_j \mid A_i)P(A_i)}{P(B_j \mid A_1)P(A_1) + P(B_j \mid A_2)P(A_2) + \ldots + P(B_j \mid A_k)P(A_k)}.$$

(4.22)

Example 4.7

A police officer stops a speeding driver on a Saturday night and gives him a Breathalyzer test. The test shows that his estimated blood alcohol content is above .08 g/dl, and this is considered evidence of driving while intoxicated. Suppose that a Breathalyzer test has a false positive rate of 5% and a false negative rate of 10%. Also, assume that in this town on a Saturday night 8% of drivers are intoxicated. What is a chance that a driver who has a positive Breathalyzer test is actually driving while intoxicated (DWI)?

Let's let A_1 = DWI and A_2 = not DWI, and let B_1 = failed Breathalyzer test and B_2 = passed Breathalyzer test. We read above that $P(A_1) = .08$ and calculate that $P(A_2) = 1 - P(A_1) = .92$. We are given that $P(B_1 \mid A_1) = .90$ and $P(B_1 \mid A_2) = .05$. Using Bayes' theorem we get:

$$P(A_1 \mid B_1) = \frac{.90 \times .08}{.90 \times .08 + .05 \times .92} \approx .61.$$

Thus, the chance that someone with a positive Breathalyzer test is actually driving while intoxicated is about 0.61. In addition, the probability of correctly identifying a DWI depends upon the percentage of people in the population who are driving while intoxicated. For example, suppose that no one was driving under the influence of alcohol. If one were to ask "What is the probability that someone driving under the influence would have a positive Breathalyzer test?" the answer clearly would be zero.

4.10 CONCLUDING REMARKS

Transportation engineers are often interested in phenomena that are not deterministic. In this situation they make use of random variables. Random variables can take on a range of values, either discrete or continuous, with a distinct probability or likelihood associated with each value or range of values. In these situations, the easiest way to discuss these phenomena is through mathematics, since each phenomenon is defined as a function or random variable. For continuous distributions the function is known as a probability density function (pdf), and the probability is represented by the area under the function. In other words, the transportation professional will have to integrate the function over the range of values of interest. For discrete variables the function is known as the probability mass function (pmf), where it represents the probability of a given event. It will be shown in the following chapters that some functions are used more than others, and these will be assigned distinct names. However, the principles underlying pdfs and pmfs are universal.

Sometimes the engineer is not interested in the distribution per se, but in certain characteristics of the distribution. Common metrics include measures of central tendency (e.g., mean, mode, median) and measures of dispersion (e.g., standard deviation). Not surprisingly, engineers are also interested in the relationship between different random variables; for example, roadway density is a function of speed and flow rate. Because both of the latter variables are random, then so density also is a random variable. As was demonstrated, it is often difficult to identify the exact form of the pdf of a random variable that is a function of two or more random variables except under special circumstances. However, it is often easier to calculate or approximate the expectation of this random variable (e.g., mean, variance). For example, one might not be able to derive the pdf of the roadway density given the pdfs of speed and flow, but one could potentially derive the mean and variance of the pdf of roadway density. For many applications this may be sufficient. These concepts will be discussed in greater detail later in the text.

This chapter introduced the concepts of probability and random variables. The authors assume that most readers have been exposed to these concepts before, and consequently this introduction was relatively brief. However, these concepts form the basis for the remainder of the book; hence, if a student has not mastered these basic concepts it will be nearly impossible to successfully master the material that follows. We encourage readers to follow up with some of the supplementary material if the concepts of probability, random variables, pdfs, and pmfs are not clear.

HOMEWORK PROBLEMS

1. What is the sample space for the following?

    a. A state legislature is considering lowering freeway speeds to reduce fatal accidents.

b. A city planning commission is considering recommending that bicycle paths be constructed. They want citizen input. What would the sample space be for the survey?

2. Suppose that the odds of a person dying in a plane crash are 1/9,000,000 and the odds of surviving a plane crash are 50%. Explain how both probabilities can be simultaneously accurate.

3. Suppose that the probability that a sports car driver gets a speeding ticket during a year is 5%. Suppose that 8% of male sports car drivers get tickets during a year. Suppose that 55% of drivers are female. What is the probability that a female sports car driver gets a ticket in a year?

4. A random variable U is uniformly distributed in the interval $[-1,1]$. That means that its pdf is

$$f(u) = \begin{cases} \dfrac{1}{2}, -1 \le u \le 1 \\ 0 \text{ elsewhere} \end{cases}.$$

Calculate the mean and variance for U.

5. This question is known as the birthday problem: Assume that there are 365 days in a year (ruling out leap years) and that the same number of people are born on each day of the year. How many people have to be (randomly arrive) in a room before the chance of having two people with the same birthday is at least 1/2?

6. The variance of a random variable has to be greater than or equal to zero. It equals zero if and only if the random variable is a constant with probability 1. The kurtosis of a random variable X is often defined as

$$\frac{E(X^4)}{\sigma^4} - 3.$$

It is a measure of the likelihood of outliers for the random variable X. In general, the bigger the kurtosis is, the more likely outliers will occur. Show that the uniform distribution defined in problem 4 has minimum kurtosis. (Hint: What is the variance of U^2?)

7. When working with transportation statistics that cover several orders of magnitude, it is common to use a log transform to plot the data. Otherwise, the plot is often hard to understand. Using the mobility analysis data, calculate the coefficient of variation for freeway miles and calculate one hundred times the standard deviation of log freeway miles. Are the statistics close? Why or why not?

8. A police officer stops a speeding driver on a Saturday night and gives the driver a Breathalyzer test. The test shows that the person's blood alcohol content is above .01, and that is evidence of driving while intoxicated. Assume that a Breathalyzer test has a false positive rate of 3% and a false negative rate of 5%. Also, estimate that on a Saturday night in this town 10% of drivers are driving while intoxicated. What are the chances who a driver who has a positive Breathalyzer test is actually driving while intoxicated?

REFERENCES

Devore, J. L. 2008. *Probability and statistics for engineering and the sciences.* 7th ed. Pacific Grove, CA: Brooks/Cole Publishing Co.

Ku, H. H. 1966. *Journal of Research of the NBS.* Notes on the use of error formulas. 70C.4:263–73.

May, A. D. 1990. *Traffic flow fundamentals.* Saddle River, NJ: Prentice Hall.

Ott, R., and M. Longnecker. 2009. *An introduction to statistical methods and data analysis.* 6th ed. Belmont, CA: Cengage/Brooks/Cole.

CHAPTER 5

Common Probability Distributions

5.1 INTRODUCTION

Inexperienced transportation professionals sometimes make the mistake of thinking that all variables are deterministic; that is, they have a known and constant value. For example, in most introductory classes engineers are exposed to the fundamental equation of traffic flow, which relates flow (vehicles/time unit), q, to density (vehicles/distance unit), d, and velocity (distance unit/time unit), v. That is,

$$q = dv.$$

However, if one were to go measure any of these parameters in the field it would quickly become obvious that there is considerable variability in the observations. This is true of almost all engineering equations used in practice. This does not mean the equations are incorrect. Rather, it may be that the equation refers to average conditions, or it may indicate that the formulas are a useful approximation. The most famous example of this is Newton's law that force equals the product of mass and acceleration. Einstein showed that, in general, this model is incorrect. Nonetheless, it is easy to argue that the model is extremely useful in bridge design.

In those situations where it is not productive to treat variables as deterministic, engineers employ probability theory to model the variables. These are known as random variables, as defined in Chapter 4, and they have a probability distribution. Probability mass functions (pmfs) are used for discrete variables, and probability density functions (pdfs) are used for continuous variables. Many of the more useful functions are used so often that they are given special names. These common functions, and their properties, will form the focus of this chapter.

It is important to note that many engineering formulas, which from all outward appearances are deterministic, actually include these probabilistic concepts. For example, most design codes (e.g., roadway, pavement mixture, bridge, etc.) are developed with probabilistic theory and then written in a manner that appears deterministic to an unsophisticated reader. For example, the rules on reinforcing bar size in pavement design manuals are deterministic even though the recommended values are based on statistical theory. A conventional method of accounting for variability is the safety factor that is inherent in many designs. The commonly accepted reason for this is that most engineers would not be able to understand the probability and statistics concepts if they were written in a more mathematically rigorous manner.

5.2 DISCRETE DISTRIBUTIONS

A random variable X is referred to as a discrete random variable when the set of all possible outcomes (sample space) is countable. This does not mean that one would ever count the outcomes; rather, it simply refers to whether it would be possible to count them. While there are many pmfs, there are only a few discrete distributions that are often used in transportation. For example, the number of times that a new video imaging vehicle detection system detects a vehicle correctly, out of n passing vehicles, can be modeled by a binomial distribution. Meanwhile, pedestrian arrivals at a crosswalk during the given time period may be modeled by a Poisson distribution. Similarly, crash counts at an intersection may be modeled by a Poisson distribution or a negative binomial distribution. We briefly introduce some of those distributions in this chapter, but for more extensive discussions on discrete distributions, see the book by Casella and Berger (1990). Additionally, for a complete treatment of the various families of discrete distributions, please refer to Johnson and Kotz (1969).

5.2.1 Binomial Distribution

The binomial distribution is used to model the number of successes in n independent trials where each trial has a binary outcome: either success (when an event occurs) or failure (when an event does not occur). Let p be the probability of success. Taking integer values between 0 and n, a random variable has a binomial distribution if its probability mass function (pmf) is given by

$$P(X=x)=\binom{n}{x}p^x(1-p)^{n-x},$$

(5.1)

for $x=0,1,2,\dots,n$. The binomial distribution is often denoted as Bin (n, p), where n is the number of trials and p is the probability success. We get different binomial distributions with different n and p. Figure 5.1 illustrates three binomial distributions (pmf): (a) with $n = 10$ and $p = 0.2$, (b) with $n = 10$ and $p = 0.5$, and (c) with $n = 10$ and $p = 0.8$. It can be observed from the plots for fixed n that as p gets close to 0.5, the distribution becomes more symmetric, and the distribution is symmetric at $p = 0.5$. It should be noted that each outcome (say when $n = 10$

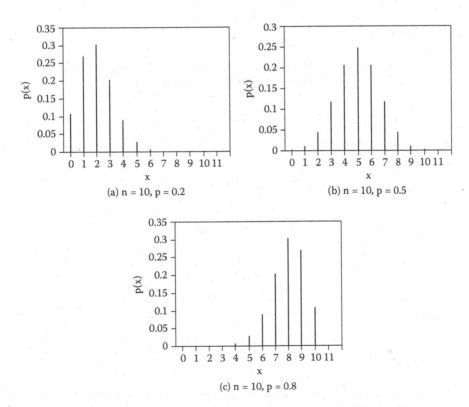

FIGURE 5.1 Binomial distributions with (a) $n = 10$, $p = 0.2$, (b) $n = 10$, $p = 0.5$, and (c) $n = 10$, $p = 0.8$.

and $x = 9$ in Figure 5.1a) has a nonzero probability associated with it. However, some of the probabilities are so small they cannot be viewed on the graphs.

It can be shown (see, e.g., Examples 2.2.2 and 2.3.2 of Casella and Berger, 1990) that the mean and variance of the binomial distribution are given as

$$E(X) = \sum_{x=0}^{n} x \binom{n}{x} p^x (1-p)^{n-x} = np \tag{5.2}$$

and

$$Var(X) = \sum_{x=0}^{n} x^2 \binom{n}{x} p^x (1-p)^{n-x} - (np)^2 = np(1-p). \tag{5.3}$$

Remark 5.1

Each trial underlying the binomial distribution is called a Bernoulli trial. A random variable X is said to have a Bernoulli (p) distribution if

$$P(x = 1) = p \quad \text{and} \quad P(x = 0) = 1-p \tag{5.4}$$

which can also be written as

$$P(X = x) = p^x(1-p)^{1-x}, \, x = 0,1. \tag{5.5}$$

The binomial random variable $Y \sim \text{Bin}(n, p)$ can be regarded as the sum of n independent Bernoulli random variables; that is,

$$Y = \sum_{i=1}^{n} X_i$$

where $X_i \sim Bernoulli\ (p)$.

Example 5.1

The manufacturer of a new video imaging vehicle detection system advertises that the nondetection rate (the probability of not detecting a vehicle when there is a vehicle present) is known to be 0.05. What is the probability that more than two out of the next ten vehicles that pass the imaging system are not detected?

In this case, the expected number of vehicles that would not be detected, out of the ten that pass the site, can be modeled by a binomial distribution composed of ten trials and a nondetection rate of 0.05 as the success probability: $p = 0.05$. Let X be the number of vehicles that are not detected by the video system in the ten passing vehicles. Then we have the following:

$$P(X > 2) = 1 - P(X \le 2)$$

$$= 1 - \binom{10}{0}(0.05)^0(0.95)^{10} - \binom{10}{1}(0.05)^1(0.95)^9$$

$$- \binom{10}{2}(0.05)^2(0.95)^8$$

$$= 1 - 0.5987 - 0.3151 - 0.0746 = 0.0116.$$

There is only a 1.16% chance that more than two out of ten vehicles are not detected by this imaging system. It should be noted that this assumes the manufacturer is correct in its assertion. Methods for measuring these types of claims will be addressed in Chapter 9.

5.2.2 Poisson Distribution

A Poisson distribution is used to model the number of occurrences (or arrivals) in a given time interval (or space) when it can be assumed that the probability of observing an occurrence does not depend on time and is proportional to the waiting time (if the waiting time is small). See Casella and Berger (1990) for a set of rigorous assumptions underlying the Poisson distribution. The mean of a Poisson distribution (λ) is assumed proportional to the length of time. If t denotes the length of time for an observation period, then typically $\lambda = \lambda_0 t$, where λ_0 is the mean of the Poisson for one observation time unit. A random variable, taking nonnegative integer values, has a Poisson distribution, denoted by Poisson (λ), if its probability mass function (pmf) is given by

$$P(X = x) = \frac{\exp(-\lambda)\lambda^x}{x!}, \tag{5.6}$$

for $x = 0,1,2,\ldots$. As can be seen above, it is completely determined by a single parameter, λ, which is both the mean and the variance of the Poisson distribution. Figure 5.2 presents plots for two Poisson pmfs: one with $\lambda = 1$ and the other with $\lambda = 5$. It can be observed from the

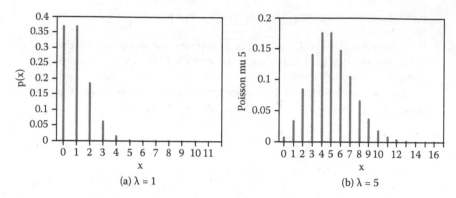

FIGURE 5.2   Poisson distributions with (a) $\lambda = 1$ and (b) $\lambda = 5$.

plots that as λ gets larger, all else being equal, the distribution becomes more symmetric.

Example 5.2

The arrival rate at a pedestrian crosswalk is known to be 0.2 arrival per minute—in other words, a pedestrian arrives every five minutes on average. What is the probability that there will be no pedestrian arrivals in the next minute? What is the probability there will be more than two arrivals?

In this case, the number of pedestrian arrivals per minute at a crossing can be modeled by a Poisson distribution having an arrival rate of 0.2 (i.e., $\lambda = 0.2$). Let X be the number of arrivals in a minute. Then we have the following:

$$P(X=0) = \frac{\exp(-0.2)0.2^0}{0!} = \exp(-0.2) = 0.8187.$$

There is about an 82% chance that no pedestrian arrives at a crossing in the next minute. Also,

$$P(X>2) = 1 - P(X \le 2)$$

$$= 1 - \frac{\exp(-0.2)0.2^0}{0!} - \frac{\exp(-0.2)0.2^1}{1!} - \frac{\exp(-0.2)0.2^2}{2!}$$

$$= 1 - 0.8187 - 0.1637 - 0.0164$$

$$= 1 - 0.9989$$

$$= 0.0011.$$

There is only an 0.11% chance that more than two pedestrian arrivals will be observed in the next minute.

5.2.3 Negative Binomial Distribution

In probability and statistics, the negative binomial distribution is often used to model the number of failures observed before a fixed number of successes (say, r) in independent Bernoulli trials. Let p be the probability of success. A random variable, Y, taking nonnegative integer values has a negative binomial distribution denoted by NB (r,p) if its pmf is given by

$$P(Y=y)=\frac{\Gamma(r+y)}{\Gamma(r)\Gamma(y+1)}p^{r}(1-p)^{y} \qquad (5.7)$$

$y = 0,1,\ldots$, where $\Gamma(x)$ is a gamma function. The mean and variance can be shown to be

$$E(Y)=\sum_{y=0}^{\infty}y\frac{\Gamma(r+y)}{\Gamma(r)\Gamma(y+1)}p^{r}(1-p)^{y}=\frac{r(1-p)}{p} \qquad (5.8)$$

and

$$Var(Y)=\sum_{y=0}^{\infty}y^{2}\frac{\Gamma(r+y)}{\Gamma(r)\Gamma(y+1)}p^{r}(1-p)^{y}-r^{2}\frac{(1-p)^{2}}{p^{2}}=\frac{r(1-p)}{p^{2}}, \qquad (5.9)$$

respectively (see, e.g., Casella and Berger, 1990).

We get different negative binomial distributions with different r and p. Figure 5.3 presents examples of negative binomial distributions (pmf) for different values of r and p.

The negative binomial distribution may also be defined in terms of the number of trials, X, until the rth success is observed. Using the relationship that $X = r + Y$, the pmf for the number of trials can be given as follows (see, e.g., Casella and Berger, 1990):

$$P(X=x)=\frac{\Gamma(x)}{\Gamma(r)\Gamma(x-r+1)}p^{r}(1-p)^{x-r} \qquad (5.10)$$

for $x = r, r + 1,\ldots$.

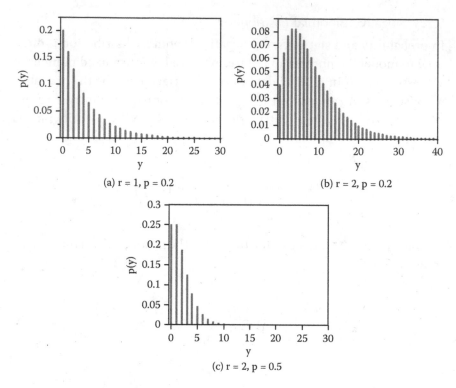

FIGURE 5.3   Negative binomial distribution with (a) $r = 1$, $p = 0.2$, (b) $r = 2$, $p = 0.2$, and (c) $r = 2$, $p = 0.5$.

Remark 5.2

The negative binomial distribution with $r = 1$ is specifically called a geometric distribution, denoted by geometric (P). It can be used to model the number of trials until the first success is observed. The pmf can be written as

$$P(X = x) = p(1-p)^{x-1} \tag{5.11}$$

for $x = 1,2,\ldots,$ with the mean and variance given by $E(X) = (1/p)$ and $Var(X) = (1-p)/p^2$, respectively.

So far, the negative binomial distribution is defined using the parameters p and r. Other parameterizations of the negative binomial distribution are also available. One such parameterization that is often used in transportation, especially in safety analysis, is given in terms of its mean. Let us denote the mean of the negative binomial distribution given in Equation 5.8 by λ. Thus,

$$\lambda = \frac{r(1-p)}{p}. \tag{5.12}$$

Then it can be shown that

$$Var(Y) = \frac{r(1-p)}{p^2} = \frac{1}{r}\lambda^2 + \lambda. \tag{5.13}$$

Because $(1/r)\lambda^2$ is positive, $Var(Y)$ is larger than $E(Y)$, and the negative binomial distribution can be used in place of the Poisson distribution to model the count data when the observed variance of data is greater than the mean (the phenomenon referred to as overdispersion). In fact, the Poisson distribution can be viewed as a limiting case of the negative binomial distribution: $r \to \infty$ and $p \to 1$ such that $r(1-p) \to \lambda$ (see, e.g., Casella and Berger, 1990). Under the parameterization given in Equation 5.12, the negative binomial distribution in Equation 5.7 can be redefined as

$$P(Y = y) = \frac{\Gamma(r+y)}{\Gamma(r)\Gamma(y+1)}\left(\frac{r}{\lambda+r}\right)^r\left(\frac{\lambda}{\lambda+r}\right)^y. \tag{5.14}$$

If r is also replaced by k^{-1} in Equation 5.14, then we have the form of the negative binomial distribution that is often used in safety analysis to model the number of crashes, namely,

$$P(Y = y) = \frac{\Gamma(k^{-1}+y)}{\Gamma(k^{-1})\Gamma(y+1)}\left(\frac{k^{-1}}{\lambda+k^{-1}}\right)^{1/k}\left(\frac{\lambda}{\lambda+k^{-1}}\right)^y \tag{5.15}$$

$y = 0,1,2,\dots$. The parameter k is often referred to as the negative binomial dispersion parameter. The examples of the negative binomial distribution given in the form of Equation 5.15 can be found in Chapter 11.

5.3 CONTINUOUS DISTRIBUTIONS

When the values of the variable can theoretically take on any value on the real number line or an interval, the variable is said to be continuous (see Chapter 4). For example, traffic engineers have found it useful to model variables—such as travel times, speeds, and stopping distances—as having continuous distributions. Many distributions used in transportation are nonnegative because most physical phenomena that engineers are interested in do not have negative values. For example, the lognormal distribution is used to model forces in bridge engineering and it does not allow negative values. As an aside, sometimes transportation engineers use pdfs that allow negative values even when the phenomenon of interest is nonnegative. For example, a transportation professional may find it convenient to model link travel times with a normal pdf. However, in these situations the probability of a negative value typically is very small and, as a result, can be ignored. The important point is that all models will have error, and the engineer must discern whether the information provided by the model is useful.

Note that in practice the empirical values of these variables may be rounded to decimal units. For example, speeds are usually recorded to only a few decimal points (depending on the detector technology used), but they are frequently modeled as coming from a continuous distribution. In practice, transportation engineers tend to use only a few of the available continuous pdfs. These commonly used distributions require special attention and are listed below.

5.3.1 Normal Distribution

Perhaps the most commonly used distribution by transportation engineers is the normal distribution. This is sometimes referred to as the Gaussian distribution, and is more commonly known to students as the bell curve. This distribution is popular for three reasons. First, it has a central role in transportation statistics. As described by the central limit theorem (CLT) in Chapter 6, the averages of statistically independent measurements have a probability distribution that closely approximates a normal distribution. When estimates are averages of large ensembles they will often be approximately normally distributed. Secondly, the normal distribution is used as the motivation for most statistical tests (such as the t-test) so that those tests are typically appropriate when applied to average speeds and travel time indices. Finally, the normal pdf is easy to optimize and is a commonly used model for errors and survey measurements. Therefore, while many traffic phenomena are the result of myriad inputs, if the results are additive, then the resulting random variable will be normally distributed. For example, there are many factors that affect walking speed. However, because these effects are both additive and independent, it can be readily shown empirically that walking speed may be modeled by the normal distribution.

A random variable is said to have a normal distribution with mean μ (center and mode of the normal distribution) and variance σ^2, if its pdf has the form

$$f(x;\mu,\sigma^2)=\frac{1}{\sqrt{2\pi}\,\sigma}e^{-\frac{(x-\mu)^2}{2\sigma^2}}.$$

(5.16)

Note: The normal distribution is often written in short form as $N(\mu,\sigma^2)$. For example, $N(10, 2)$ would imply that a variable is distributed normally with a mean of ten units and a variance of two units. However, some authors use the short form $N(\mu,\sigma)$. Therefore, it is critical that the engineer understands the format a particular author is using so he or she knows whether the second parameter is variance or standard deviation.

Both the variance and standard deviation, σ, are indicators of spread. For the normal distribution, the latter is more often used as a measure of spread because of some special structure inherent in the pdf, which will be discussed later. Also, the variance is given in squared units and is often awkward to use directly. The larger σ is, the more spread out the distribution will be. The parameters of the normal distribution, μ and σ, are frequently estimated by the sample mean $(\bar{X})$ and sample standard deviation (S). A normal distribution with $\mu = 1$ and $\sigma = 1$ is known as the standard normal distribution, denoted by $N(0, 1)$. Figure 5.4 shows the pdf of the standard normal distribution.

Note that there are no closed-form solutions for finding the area (e.g., calculating the probability) under the normal pdf. The integration has to be done using numerical methods. However, for most of the common distributions, these are already programmed for engineers. For example, Excel™ has a function for calculating the probability for any normal distributions as long as the mean, variance, and range of interest are known.

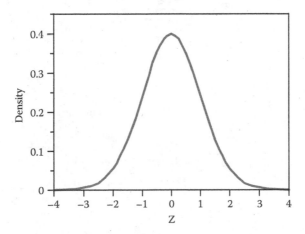

FIGURE 5.4   Density of standard normal distribution.

> **Note:** Prior to the advent of computers it was common to use tables to look up the values of the normal distribution. Needless to say, it would be impossible to develop the tables for all combinations of means and variances. Instead, statisticians would transform a given distribution to an equivalent normal distribution with a mean of 1 and a standard deviation of 1—the standard normal distribution. Many textbooks still use this approach, but we consider this method a bit archaic given the computer and software technology now available to students and researchers. Readers can refer to Ott and Longnecker (2009) for a method for finding the normal percentiles from a table.

Normal percentiles (quantiles) can easily be computed by JMP. In order to demonstrate how to obtain quantiles from the normal distribution using JMP, we start by assuming that we are interested in the standard normal distribution. If we want to get only one percentile, we can add a column to the data and choose the formula from the pull-down menu under "Cols" in the JMP menu bar. Then, in the formula menu we can choose "Normal Quantile" as shown in Figure 5.5, and insert the percentage that we want to use and choose "apply." Otherwise, for a set of percentiles, we first enter the percentages that we are interested in into a JMP data table column. For example, we enter 0.01, 0.05, 0.1, 0.9, 0.95, and 0.99 in column

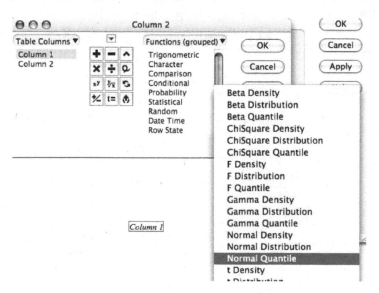

FIGURE 5.5   Formula menu from JMP.

Column 1	Column 2
0.01	−2.3263479
0.025	−1.959964
0.05	−1.6448536
0.1	−1.28155157
0.9	1.281585157
0.95	1.64485363
0.975	1.95996398
0.99	2.32634787

FIGURE 5.6   Output of standard normal percentiles obtained by JMP.

1 (see column 1 of Figure 5.6). Then we create a new column, say column 2, and follow the sequence shown in Figure 5.5. The output from this command is shown in Figure 5.6. The percentiles that we want are in column 2. See "Probability Functions" from JMP Help for more explanations on how to obtain probabilities and quantiles for normal distributions.

For example, the 97.5th percentile from a standard normal distribution is approximately 1.96—a familiar value to many transportation professionals. In other words, 95% of the values in any normal distribution will lie between plus and minus 1.96 standard deviations from the mean (e.g., each tail will hold 2.5% of the probability).

So far we have been exploring the standard normal distribution and have shown how to obtain percentiles from it. Suppose that X has a normal distribution with mean μ and variance σ^2. Then a standard normal random variable, typically denoted by Z, can be obtained from X using the equations

$$Z = \frac{X-\mu}{\sigma} \tag{5.17}$$

and

$$X = \mu + \sigma Z. \tag{5.18}$$

Thus, if we know the percentiles from the standard normal distribution, the percentiles for any other normal distributions can also be obtained by plugging in the values in Equation 5.18.

Example 5.3

Say that X has a normal distribution with mean 55 Mph and standard deviation 3 mph. Then the 2.5th percentile can be obtained by

$$55 + 3(-1.96) = 55 - 3(1.96) = 49.12$$

and the 97.5th percentile can be obtained by

$$55 + 3(1.96) = 60.88.$$

It follows that the central 95% of this normal distribution lies between 49.12 and 60.88.

5.3.2 t-Distribution

The t-distribution has a shape similar to that of the normal distribution (symmetric and bell shaped), but has heavier tails. This means that there is more area under the t-distribution pdf than under the normal pdf far from the mean. The t-distribution can be used to model averages when the original data have a large number of outliers. Recall that the gamma function is defined by the equation $\Gamma(x) = \int_0^\infty t^{x-1} e^{-t} dt$ for any positive x. The pdf corresponding to a t-distribution with ν degrees of freedom is

$$f_\nu(t) = \frac{\Gamma\left(\dfrac{\nu+1}{2}\right)}{\sqrt{\pi \nu} \, \Gamma\left(\dfrac{\nu}{2}\right)} \left(1 + \frac{t^2}{\nu}\right)^{-(\nu+1)/2}. \tag{5.19}$$

This pdf is symmetric about zero, and provided that $\nu \geq 3$, the variance of a t-distributed random variable is

$$Var(t_\nu) = \frac{\nu}{\nu - 2} \tag{5.20}$$

and the mean is zero. A t-distribution with $\nu = \infty$ is a standard normal distribution, and a t-distribution with $\nu = 1$ is a Cauchy distribution (Johnson et al., 1994) that has neither a mean nor a variance. We can obtain graphs showing t-distributions with many different degrees of freedom. Figure 5.7 shows a t-distribution with 3 degrees of freedom.

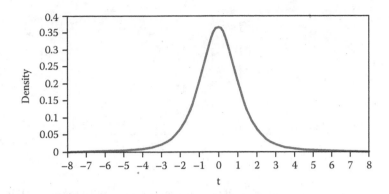

FIGURE 5.7 t-density with 3 degrees of freedom.

Percentiles for t-distributions can be obtained from JMP by referring to the JMP manual for details. The distribution of the test statistic

$$t = \frac{\bar{X} - \mu_0}{S/\sqrt{n}},$$

that will be introduced in Chapter 7 also has a t-distribution with $n - 1$ degrees of freedom.

5.3.3 Lognormal Distribution

The lognormal distribution is an asymmetric heavy-tailed distribution (skewed to the right) that is useful in many engineering studies. The obvious advantage is that the distribution is nonnegative and is heavy tailed; therefore, it has some advantages over the normal distribution (which can have values to negative infinity and a few outliers) for modeling physical phenomena. In transportation it has been used in studies of signing and visibility. For example, see Bachman et al. (2006), where a log transform was used before implementing t-tests.

Data are lognormally distributed when the log transform of the data is normally distributed. That is, for a random variable, X, if $\log(X) \sim N(\mu, \sigma^2)$, then $X \sim LN(\mu, \sigma^2)$. Its pdf is

$$f(x; \mu, \sigma^2) = \frac{1}{\sqrt{2\pi}\sigma x} e^{-\frac{[\log(x) - \mu]^2}{2\sigma^2}} \tag{5.21}$$

for nonnegative x and is zero elsewhere. Its mean is

$$E(X) = e^{\mu + \sigma^2/2} \tag{5.22}$$

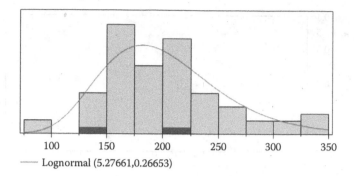

—— Lognormal (5.27661,0.26653)

FIGURE 5.8   Pavement marking detection distance data with lognormal density overlaid.

and variance is

$$Var(X) = \mu^2(e^{\sigma^2} - 1).$$
(5.23)

Note that the parameters of the lognormal distribution, μ and σ^2, are not the mean and variance, but the shape and scale parameters of the distribution. Usually the estimated mean and standard deviation for log-transformed data are reported as the shape and scale parameters, respectively. For example, the histogram in Figure 5.8 comes from a pavement marking visibility study in rainy conditions (Carlson et al., 2007). The distance from pavement marking to detection (detection distance) is the variable of interest.

We can see that the overlaying lognormal density to the histogram fits the data reasonably well. There are some people who identify the marking relatively quickly (e.g., at 100 m), but there are a relatively large proportion of people who need much longer (e.g., greater than 250 m). The empirical data are asymmetric and, in this situation, the lognormal is a reasonable model. The estimated shape parameter (μ) is 5.28 and the estimated scale parameter (σ^2) is 0.27.

5.3.4 Exponential Distribution

The exponential distribution is often used to model arrival times of vehicles and pedestrians at intersections and other points of interest. Should the number of arrivals satisfy the following assumptions, then the exponential distribution is usually an appropriate model for arrival times. These assumptions are as follows:

1. The number of arrivals can be modeled as a stationary Poisson process (the mean of the Poisson process does not change with time).

2. The number of arrivals at any particular time interval is at most one.

3. The arrivals in different time periods are independent.

The probability density function for the exponential distribution is given by

$$f(x) = \tfrac{1}{\beta} e^{-x/\beta} \tag{5.24}$$

for $x \geq 0$. The mean of the exponential distribution is β (the expected time until a single arrival), and the variance is β^2. The exponential density is considered to be heavy tailed—hence, exponential data typically exhibit many more outliers than normally distributed data. The exponential distribution is often written $EXP(\beta)$; that is, it is a one-parameter pdf in contrast to the normal and lognormal pdfs that require two parameters to define the function. As an example, consider the pedestrian data obtained at a Tucson, Arizona, site (Fitzpatrick et al., 2006) shown in Figure 5.9. The sample mean for this data is 216 and the sample standard deviation is 190. The fact that they are nearly equal is characteristic of data from an exponential distribution.

The exponential distribution is used in many transportation applications. For example, it has been used for modeling walking distance to transit stops, meaning that most people who take transit are willing to walk short distances and fewer are willing to walk long distances. Knowing this type of information is vital for designing transit systems and identifying stop locations.

5.3.5 Gamma Distribution

The gamma distribution used by transportation engineers to model headway or spacings (often measured as time) between vehicles has an

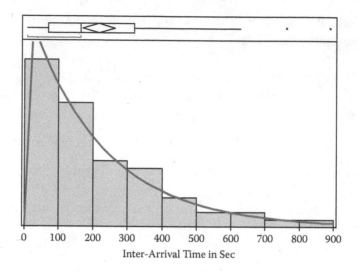

FIGURE 5.9   JMP output containing the histogram of the interarrival time data with exponential density overlaid.

extra parameter, β. The gamma distribution has a probability density function

$$f(x) = \frac{x^{\alpha-1}e^{-x/\beta}}{\Gamma(\alpha)\beta^{\alpha}} \qquad (5.25)$$

for $x \geq 0$. The mean of this more general gamma distribution is $\alpha\beta$ and its variance is $\alpha\beta^2$. The parameter α is called the shape parameter and the parameter β is called the scale parameter. The distribution is written as *Gamma*(α, β).

The gamma distribution has heavy tails as Figure 5.10 demonstrates. This example depicts headway data collected by a Texas A&M University graduate student for her transportation statistics class.

The exponential distribution is a special case of the gamma distribution where $\alpha = 1$ in the gamma distribution.

5.3.6 Chi-Square Distribution

It should be apparent to the reader that there are a number of common pdfs. Some of these—lognormal, exponential, gamma—are used to model physical phenomena, others are used almost exclusively for statistical tests, and some are used for both (e.g., normal distribution). The chi-square distribution is

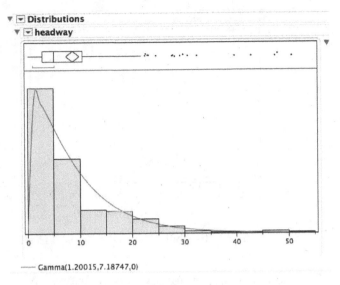

▼ ☑ **Distributions**
　▼ ☑ **headway**

Gamma(1.20015,7.18747,0)

FIGURE 5.10   Histogram of the headway data with gamma density overlaid.

used almost exclusively for statistical testing. The pdf corresponding to a chi-square distribution with v degrees of freedom is given by

$$f(x) = \frac{x^{(v/2)-1}e^{-x/2}}{\Gamma(v/2)2^{v/2}}, x \geq 0. \tag{5.26}$$

The chi-square distribution is a special case of the gamma distribution, as can be seen from the following: If $X_1, X_2, \ldots, X_n$ are independent standard normal random variables, $N(0,1)$, then $\sum_{i=1}^{n} X_i^2$ has a gamma distribution with $\alpha = n/2$ and $\beta = 2$. That distribution is also called a chi-square distribution with n degrees of freedom. The chi-square distribution is most commonly used to model the distribution of variance estimates from normally distributed data. In particular, for independent normally distributed data with mean μ and variance σ^2 the ratio

$$\frac{S^2}{\sigma^2} = \frac{\displaystyle\sum_{i=1}^{n}(X_i - \bar{X})^2/(n-1)}{\sigma^2}$$

has a chi-square distribution with $n - 1$ degrees of freedom (see Chapter 6).

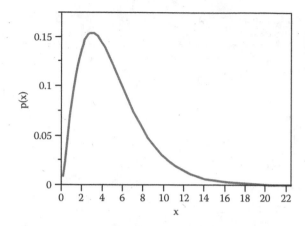

FIGURE 5.11   pdf of chi-square distribution with 5 degrees of freedom.

Figure 5.11 shows a graph of a chi-square density with 5 degrees of freedom.

5.3.7 *F*-Distribution

The *F*-distribution is encountered in many statistical hypothesis-testing situations, including ANOVA, regression, and tests for equality of variances that will be introduced in Chapter 7. The *F*-distribution arises with the distribution of the ratio of independent chi-square random variables (each divided by its degrees of freedom). That is, for two independent chi-square random variables, $\chi^2_{1,df1}$ and $\chi^2_{2,df2}$, with df1 and df2 degrees of freedom, respectively,

$$F = \frac{\chi^2_{1,df1}/(df1)}{\chi^2_{2,df2}/(df2)} \tag{5.27}$$

has an *F*-distribution with *df*1 and *df*2 degrees of freedom. The pdf for the *F*-distribution is complicated and infrequently used by engineers. The pdf can be found by referencing Johnson et al. (1994). These ratios commonly occur as ratios of variance estimates based upon normally distributed data. The order for the degrees of freedom for the *F*-distribution is relevant and cannot be interchanged. For example, Figure 5.12 shows *F*-densities with (a) *df*1 = 2 and *df*2 = 5 and (b) *df*1 = 5 and *df*2 = 2 degrees of freedom, respectively.

Percentiles for *F*-distributions can be obtained from JMP, as in the case of normal percentiles (see the JMP manual for details). For example, the

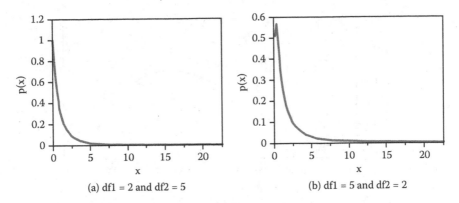

(a) df1 = 2 and df2 = 5   (b) df1 = 5 and df2 = 2

FIGURE 5.12   *F*-densities with (a) *df1* = 2, *df2* = 5 and (b) *df1* = 5, *df2* = 2.

95th percentile for an *F*-distribution with 2 and 5 degrees of freedom can be obtained as 5.786.

5.3.8 Beta Distribution

The parent of the *F*-distribution is the beta distribution. Besides parenting the *F*-distribution the beta distribution is a very flexible distribution for modeling random variables that are restricted to the interval of 0 to 1. Its pdf is

$$f(x;\alpha,\beta)=\frac{\Gamma(\alpha+\beta)}{\Gamma(\alpha)\Gamma(\beta)}x^{\alpha-1}(1-x)^{\beta-1}. \qquad (5.28)$$

It can be seen that when both α and β are equal to 1, the uniform distribution is a special case of the beta distribution. In Chapter 14 we will discuss Bayesian estimation. The beta distribution will be used as the (conjugate) prior distribution for the binomial distribution. Different values of α and β give rise to differently shaped pdfs, as can be seen from Figure 5.13.

5.4 CONCLUDING REMARKS

It is the authors' experience that students often have trouble with the concept of probability distributions. The easiest way to understand them is by realizing that they provide an easy way to describe the frequency of outcomes. If the outcomes we are interested in are integers—number of cars, number of people—then a pmf is used. The probability that a particular outcome *x* will occur is simply the value of the pmf for that value

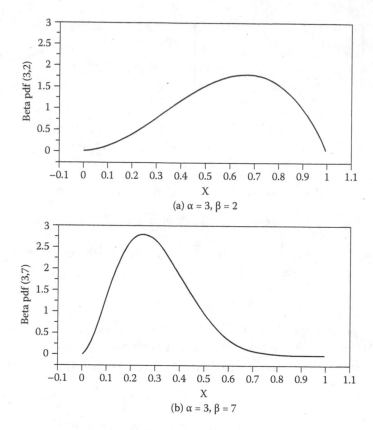

FIGURE 5.13   Beta densities with (a) $\alpha = 3$, $\beta = 2$ and (b) $\alpha = 3$, $\beta = 7$.

x. If the outcomes are continuous, then a pmf is used. In this situation the probability of the particular event x occurring is the area under the curve. Because we need to identify an area, the events are written as ranges of outcomes. For example, if X is speed of vehicles, the probability that a given vehicle will be speeding is simply the area under the curve for all values x greater than the speed limit. Note that if we were interested in the probability of a vehicle going exactly 60 mph the probability would be zero. If we meant instead that the speed was around 60 mph, then you would have to calculate the probability of being between 59 and 61 mph (or 55 and 65).

As described earlier, that there are a number of common pmfs and pdfs used by transportation professionals. Note that some of these—lognormal, exponential, gamma—are used to model physical phenomena; others are used almost exclusively for statistical tests (e.g., F-distribution); and some are used for both normal and t-distributions.

APPENDIX: TABLE OF THE MOST POPULAR DISTRIBUTIONS IN TRANSPORTATION ENGINEERING

Distribution	Short Form	Typical Formula	Mean	Variance
Binomial	Bin (n, p)	$P(X=x)=\binom{n}{x}p^x(1-p)^{n-x}$	np	$np(1-p)$
Bernoulli	Bernoulli (p)	$P(X=x)=p^x(1-p)^{1-x}$	p	$p(1-p)$
Poisson	Poisson (λ)	$P(X=x)=\dfrac{\exp(-\lambda)\lambda^x}{x!}$	λ	λ
Negative binomial	NB (r, p)	$P(Y=y)=\dfrac{\Gamma(r+y)}{\Gamma(r)\Gamma(y+1)}p^r(1-p)^y$	$\dfrac{r(1-p)}{p}$	$\dfrac{r(1-p)}{p^2}$
Geometric	Geometric (p)	$P(X=x)=p(1-p)^{x-1}$	$\dfrac{1}{p}$	$\dfrac{(1-p)}{p^2}$
Normal	N (μ, σ^2)	$f(x;\mu,\sigma^2)=\dfrac{1}{\sqrt{2\pi}\sigma}e^{-\frac{1}{2}\left(\frac{x-\mu}{\sigma}\right)^2}$	μ	σ^2
t-Distribution	t_ν	$f_\nu(t)=\dfrac{\Gamma\left(\frac{\nu+1}{2}\right)}{\sqrt{\pi\nu}\,\Gamma\left(\frac{\nu}{2}\right)}\left(1+\dfrac{t^2}{\nu}\right)^{-(\nu+1)/2}$	0 when ν (degrees of freedom) ≥ 2	$\dfrac{\nu}{\nu-2}$ when $\nu \geq 3$
Lognormal distribution	LN (μ, σ^2)	$f(x;\mu,\sigma^2)=\dfrac{1}{\sqrt{2\pi}\sigma x}e^{-[\ln(x)-\mu]^2/(2\sigma^2)}$	$e^{\mu+\sigma^2/2}$	$e^{2\mu+\sigma^2}\left(e^{\sigma^2}-1\right)$
Exponential	EXP (β)	$f(x)=\frac{1}{\beta}e^{-x/\beta}$	β	β^2
Two-parameter gamma	Gamma (α, β)	$f(x)=\dfrac{x^{\alpha-1}e^{-x/\beta}}{\Gamma(\alpha)\beta^\alpha}$	$\alpha\beta$	$\alpha\beta^2$
Chi-square	χ_ν^2	$f(x)=\dfrac{x^{(\nu/2)-1}e^{-x/2}}{\Gamma(\nu/2)2^{\nu/2}},x\geq 0$	ν (degrees of freedom)	2ν
Beta	Beta (α, β)	$f(x;\alpha,\beta)=\dfrac{\Gamma(\alpha+\beta)}{\Gamma(\alpha)\Gamma(\beta)}x^{\alpha-1}(1-x)^{\beta-1}$	$\dfrac{\alpha}{\alpha+\beta}$	$\dfrac{\alpha\beta}{(\alpha+\beta)^2(\alpha+\beta+1)}$

HOMEWORK PROBLEMS

1. It is known that a violation rate of red-light running at a particular intersection is 0.1. What is the probability that more than two out of twenty entering vehicles run red lights?

2. In problem 1, how would you model the number of entering vehicles at an intersection until the first red-light violation is observed? What is the probability that the number of entering vehicles (until the first red light runner is observed) is 5 or less?

3. The mean number of crashes per month at an intersection is known to be 1. What is the probability that there will not be a crash in the next month? What is the probability there will be more than two crashes?

4. For a normal distribution with mean equal to 55 and standard deviation equal to 5, find the 5, 25, 50, 75, and 95% points (percentiles).

5. For t-distributions with 1, 2, 5, 25, and 50 degrees of freedom, find the 5 and 95% points.

6. A random variable U is said to have a standard uniform distribution if its pdf is

$$f(u) = \begin{cases} 1, & \text{if } 0 \leq u \leq 1 \\ 0, & \text{elsewhere.} \end{cases}$$

Calculate the mean and variance for a standard uniform distribution. Find its 5 and 95% points.

7. Find the 5 and 95% points for a lognormal distribution with parameters $\mu = 2$ and $\sigma^2 = 8$. (Hint: Find the percent points for a normal random variable X and use the fact that a lognormal random variable can be expressed as e^X.)

8. For chi-square distributions with 1, 10, and 30 degrees of freedom, find the 1, 5, 95, and 99% points.

9. For F-distributions with (1, 2) and (2, 1) degrees of freedom, find the 5 and 95% points.

REFERENCES

Bachman, W. G., T. A. Wingert, and C. J. Bassi. 2006. Driver contrast sensitivity and reaction times as measured through a salt-covered windshield. *Optometry* 77:67–70.

Carlson, P. J., J. D. Miles, A. M. Pike, and E. S. Park. 2007. *Evaluation of wet-weather and contrast pavement marking applications: Final report*. Research Report 0-5008-2. College Station, TX: Transportation Institute.

Casella, G., and R. L. Berger. 1990. *Statistical inference*. Pacific Grove, CA: Wadsworth Brooks/Cole.

Fitzpatrick, K., S. M. Turner, M. A. Brewer, P. J. Carlson, B. R. Ullman, N. D. Trout, E. S. Park, J. Whitacre, N. Lalani, and D. Lord. 2006. *Improving pedestrian safety at unsignalized crossings*. TCRP Report 112/NCHRP Report 562. Transportation Research Board, Washington, D.C.

Johnson, N. L., and S. Kotz. 1969. *Distributions in statistics: Discrete distributions*. New York: Houghton Mifflin.

Johnson, N. L., S. Kotz, and N. Balakrishnan. 1994. *Continuous univariate distributions*. Vol. 1. New York: Wiley.

Ott, R., and M. Longnecker. 2009. *An introduction to statistical methods and data analysis*. 6th ed. Belmont, CA: Cengage/Brooks/Cole.

CHAPTER 6

Sampling Distributions

6.1 INTRODUCTION

In Chapters 4 and 5, we introduced the concepts of probability, random variables, and the distribution of populations. In probability, it is assumed that we know the population distribution, and our goal is to find the probability, or chance, of observing a particular event. For example, if we know that crash counts have a Poisson distribution with a known mean value (λ), we can calculate the relevant probabilities. In statistics, the characteristics or distribution of the population is often unknown, and the goal is to estimate unknown population characteristics, or the distribution based on the sample data. For example, we might have monthly crash frequency data, and from this we would like to estimate the average monthly crash frequency (λ). Once we have this information, we can use our knowledge of probability to make inferences.

Any subset of the population is called a sample. The population characteristic of interest is called a parameter. Say that we are interested in the average travel time from A to B during the morning peak hour on weekdays. If we could track all vehicles traveling from A to B, we would have the entire population of travel times and could answer questions directly. Often it is not possible to obtain the population. For this example, a sample of a number of probe vehicles traveling from A to B would be recorded. We would then use the sample data to make inferences about the population. For instance, the population mean parameter, μ (e.g., average travel time from A to B), can be estimated by the sample average travel time of the probe vehicles ($\bar{x}$), and the population variance parameter, σ^2, can be estimated by the sample variance (s^2). Note that while we can estimate the population parameters (μ, σ^2) with the sample mean and variance, respectively, we have not discussed the accuracy of these estimates. Answering this type of question will be a concern in the chapters that follow.

In the previous example, the true population average travel time (average travel time based on all vehicles traveling from A to B in the specified period) is the parameter of interest. This is, by definition, unknown. The sample average travel time is an estimate.

In fact, any quantity computed based on a sample is called an estimate, and the function resulting in the value estimate is called an estimator. For instance, the functional form of the sample mean before we plug in any values from the sample is an estimator. An estimate (describing a characteristic of a sample) is a counterpart of a parameter (describing a characteristic of a population) as a sample is a counterpart of a population.

As in the above example, an estimator (here a sample mean) is used to estimate the unknown population parameter (population mean). Because a sample is only part of the population, an observed (computed) value of the estimator will not be exactly the same as the value of the parameter (although we hope that they are close). More importantly, if a different sample is chosen, the observed value of the estimator will also change. That is, the estimate varies from sample to sample. Consequently, an estimator is a random variable, and we will make use of this fact to help quantify the accuracy of a given estimator. Because the value of an estimator will vary for different samples, they are often referred to as a sampling distribution. However, as this only refers to how its values are obtained, all the properties of random variables that were described in Chapters 4 and 5 still apply.

6.2 RANDOM SAMPLING

Clearly, samples from a population of interest can be selected in many different ways. Intuitively, the sample should be representative of the population if we are going to use this information to draw a conclusion about the population. One of the easiest ways of selecting a sample is to choose individuals or items that are easily accessible to a collector based on convenience. This type of sample is called a convenience sample. Unfortunately, the convenience sample is subject to an unknown bias that cannot be quantified. That is, when a value of estimator is obtained based on a convenience sample, one cannot quantify how far the estimate is from the true parameter value, which hinders generalizing any conclusions beyond the observed sample data. For example, if probe vehicles are dispatched when it is convenient to the drivers or the data collector, there is no way of knowing how close the resulting sample average—the travel time based on the probe vehicles—is to the true unknown average travel time. The probe drivers might wish to travel during off-peak periods because they are paid by the trip and can make more trips during this time because speeds are faster. This behavior would clearly bias the results, although the size of bias will mostly be unknown. In short, convenience sampling is not a recommended way of selecting a sample when we would like to make inferences about a population.

A preferable way of selecting the sample data in statistics is random sampling, which is also called probability sampling. Probability sampling is a method such that each unit in the population has a known probability of selection, and a chance method is used to choose the specific units to be included in the sample. Samples chosen by this selection method are objective and free from unknown biases, and are representative of the population. Types of probability samples include a simple random sample, a stratified random sample, and a cluster sample. The definition for a simple random sample is given below.

> A simple random sample of size n is a sample chosen in such a way that every possible subset of size n has the same chance of being selected as a sample.

It is important to note that while the definition is straightforward, this type of sampling is sometimes challenging. For the travel time examples, the engineer should randomly assign start times for each driver during the period of interest. However, the probe-vehicle drivers would also have to be given specific instructions regarding how they should drive their cars. If the probe-vehicle drivers all tend to go faster than the average, this too would bias the results. A common way to remove this bias is to instruct the probe-vehicles drivers to (safely) follow a specific vehicle that has been chosen randomly from the traffic stream.

Simple random sampling ensures that each individual or object in the population has the same chance of being included in a sample. Random sampling is a fundamental assumption for most inferential procedures in statistics. If this assumption does not hold, then it is nearly impossible to assess the uncertainty associated with an estimate based on the sampling distribution. This textbook assumes that samples are randomly generated, unless otherwise noted. In practice, this randomization can be difficult to implement, and a considerable amount of time is often spent ensuring that this assumption is valid. Readers who are interested in more in-depth discussions of various methods of random sampling may refer to the books by Lohr (1999) and Cochran (1977).

6.3 SAMPLING DISTRIBUTION OF A SAMPLE MEAN

The sample mean introduced in Chapter 3 is an estimator because its value is computed from the sample data. Intuitively, the value of the sample mean will be different for different samples, which is called sampling variability. To illustrate this point, let us consider a small finite population consisting of five speed measurements obtained from five vehicles.

Let μ and σ^2 denote the mean (population mean) and variance (population variance) of this population given in Table 6.1, respectively. The values of μ and σ^2 are

$$\mu = 62\left(\frac{1}{5}\right) + 55\left(\frac{1}{5}\right) + 66\left(\frac{1}{5}\right) + 67\left(\frac{1}{5}\right) + 58\left(\frac{1}{5}\right) = 61.6$$

and

$$\sigma^2 = \left(62-61.6\right)^2\left(\frac{1}{5}\right) + \left(55-61.6\right)^2\left(\frac{1}{5}\right) + \left(66-61.6\right)^2\left(\frac{1}{5}\right)$$

$$+ \left(67-61.6\right)^2\left(\frac{1}{5}\right) + \left(58-61.6\right)^2\left(\frac{1}{5}\right)$$

$$= 21.04.$$

Notice that because we have sampled the entire population, in this case five vehicles, we can readily calculate the population mean and variance.

To illustrate the point, let's say that instead of sampling the entire population, we could only sample two observations from the population given in Table 6.1. Obviously, we cannot sample the same vehicle twice. This type of sampling is referred to as sampling without replacement and, unless otherwise noted, is assumed throughout the text. Considering all possibilities, there are ten different ways of selecting a sample size of 2, as

TABLE 6.1   Population of Vehicle Speeds

Vehicle Number	1	2	3	4	5
Speed: x (mph)	62	55	66	67	58
Prob(x)	1/5	1/5	1/5	1/5	1/5

TABLE 6.2   All Possible Samples of Size 2 and the Corresponding Sample Means

Sample Number (i)	Vehicles Selected	Speeds (x_1, x_2)	Sample Mean ($\bar{x}$)
1	1, 2	62, 55	58.5
2	1, 3	62, 66	64.0
3	1, 4	62, 67	64.5
4	1, 5	62, 58	60.0
5	2, 3	55, 66	60.5
6	2, 4	55, 67	61.0
7	2, 5	55, 58	56.5
8	3, 4	66, 67	66.5
9	3, 5	66, 58	62.0
10	4, 5	67, 58	62.5

TABLE 6.3   Distribution of $\bar{X}$

$\bar{x}$	56.5	58.5	60	60.5	61	62	62.5	64	64.5	66.5
Prob($\bar{x}$)	1/10	1/10	1/10	1/10	1/10	1/10	1/10	1/10	1/10	1/10

shown in Table 6.2. The sample mean, $\bar{x}$, from each possible pair is also listed.

Assuming that a sample is obtained by simple random sampling, each of the above ten samples has the same chance of being selected (i.e., 1/10). Given (1) the entire population of sample means as shown in Table 6.2, and (2) the probability associated with each sample mean, we can identify the sampling distribution of the estimator, $\bar{X}$, directly as shown in Table 6.3.

Let $\mu_{\bar{X}}$ and $\sigma_{\bar{X}}^2$ denote the mean and variance of $\bar{X}$. Again, using the method from Chapter 4, it can be shown that

$$\mu_{\bar{X}} = \sum_{i=1}^{10} \bar{x}_i \left(\frac{1}{10} \right)$$

$$= 58.5\left(\frac{1}{10}\right) + 64.0\left(\frac{1}{10}\right) + 64.5\left(\frac{1}{10}\right) + 60.0\left(\frac{1}{10}\right) + 60.5\left(\frac{1}{10}\right)$$

$$+ 61.0\left(\frac{1}{10}\right) + 56.5\left(\frac{1}{10}\right) + 66.5\left(\frac{1}{10}\right) + 62.0\left(\frac{1}{10}\right) + 62.5\left(\frac{1}{10}\right)$$

$$= 61.6$$

and

$$\sigma_{\bar{X}}^2 = \sum_{i=1}^{10}(x_i - \bar{x}_i)^2\left(\frac{1}{10}\right)$$

$$= (58.5 - 61.6)^2\left(\frac{1}{10}\right) + (64.0 - 61.6)^2\left(\frac{1}{10}\right) + (64.5 - 61.6)^2\left(\frac{1}{10}\right)$$

$$+ (60.0 - 61.6)^2\left(\frac{1}{10}\right) + (60.5 - 61.6)^2\left(\frac{1}{10}\right) + (61.0 - 61.6)^2\left(\frac{1}{10}\right)$$

$$+ (56.5 - 61.6)^2\left(\frac{1}{10}\right) + (66.5 - 61.6)^2\left(\frac{1}{10}\right) + (62.0 - 61.6)^2\left(\frac{1}{10}\right)$$

$$+ (62.5 - 61.6)^2\left(\frac{1}{10}\right)$$

$$= 7.89.$$

It can be observed from the above experiment that:

1. The value of $\bar{X}$ varies from sample to sample and is not the same as the population mean (μ).

2. The mean of the distribution of $\bar{X}$ ($\mu_{\bar{x}}$) is the same as the population mean (μ).

3. The variance of the distribution of $\bar{X}$ ($\sigma_{\bar{X}}^2$) is smaller than the population variance (σ^2).

In general, the following relationships hold between $\mu_{\bar{X}}$ and μ, and $\sigma_{\bar{X}}^2$ and σ^2, respectively:

$$\mu_{\bar{X}} = \mu. \qquad (6.1)$$

For an infinite population,

$$\sigma_{\bar{X}}^2 = \frac{\sigma^2}{n} \qquad (6.2)$$

and for a finite population,

$$\sigma_{\bar{X}}^2 = \frac{\sigma^2}{n}\left(\frac{N-n}{N-1}\right) \qquad (6.3)$$

where N is the population size and n is the sample size. The factor $(N-n)/(N-1)$ is called the finite population correction (FPC) and can be replaced with 1 as long as the population size (N) is much larger than the sample size, n, for example, if n/N is less than 5 or 10%. Because in our example the population size is 5 and the sample size is 2, the finite population correction factor cannot be ignored:

$$\frac{N-n}{N-1} = \frac{5-2}{5-1} = \frac{3}{4} = 0.75.$$

Using Equations 6.1 and 6.3, we can now obtain the mean and variance of the distribution of $\bar{X}$ if we only know the mean and variance of the population (X) as follows:

$$\mu_{\bar{X}} = \mu = 61.6$$

$$\sigma_{\bar{X}}^2 = \frac{\sigma^2}{n}\left(\frac{N-n}{N-1}\right) = \frac{21.04}{2}(0.75) = 7.89.$$

Notice that we do not actually need to know all possible samples or associated probabilities (like those given in Table 6.3) to compute compute $\mu_{\bar{X}}$ and $\sigma_{\bar{X}}^2$. We showed these values in Tables 6.1 and 6.2 to illustrate the concepts. However, this was not necessary in order to use Equations 6.1

and 6.2. Realistically speaking, for most engineering questions it would be impossible to list the entire population and identify the relative probabilities for each outcome.

Equations 6.1 to 6.3 are important properties of the sampling distribution of $\bar{X}$. They can be interpreted as explained in the boxed text.

Under repeated random sampling (i.e., if we take a random sample and compute the sample mean many times), the values of $\bar{X}$ will be centered at the population mean. The variance of $\bar{X}$ will be smaller than the population variance by a factor of n for an infinite population or by a factor of $n(N-1)/(N-n)$ for a finite population, regardless of the underlying population distribution. The larger n is, the smaller the variability in the sampling distribution of $\bar{X}$. In other words, the sample mean obtained based on a large sample will be, in general, closer to the population mean than that obtained based on a small sample.

The above relationship is a good attribute of random sampling, and one would expect that the more samples one takes, the better the resulting estimate. However, there are two important points from Equation 6.2 to keep in mind. First, this relationship is only true on average. It could be possible that a random sample of size 10 may yield a better estimate of the population mean than a random sample of 100. However, there is no way to know which estimate is closer to the true population mean—obviously, if we knew the true population mean, we would not be trying to estimate it. Although we know that, on average, the sample mean based on one hundred will be closer, it is not true of all cases all of the time. Second, while the sampling distribution variance decreases with sample size (and hence the probability that a given estimate will be close to the true mean increases), this increase is at a decreasing rate. In other words, the marginal benefit of taking an additional ten samples after taking ten original samples will be greater than taking an additional ten samples after taking one hundred original samples.

While the above example was useful for illustrating the general concepts, in practice we will only take a single random sample of size n (rather than obtaining many random samples or size n) from a population. The value of the sample mean based upon the sample data is that it will be used as an estimate for the unknown population mean. It was shown that due to sampling variability, the value of the sample mean will not be the same as the population mean. Consequently, it would be useful to quantify how close the observed sample mean value is to the true unknown population mean (i.e., the accuracy of the sample mean). One way of quantifying this accuracy is to provide the probability statement. For example, what is the probability that the mean of a random sample of size n ($\bar{X}$) is within c (where c is a margin of error that can be tolerated) of the population mean? If we know the entire distribution of $\bar{X}$ (which we did for the example problem, as shown in Table 6.3), this question can be answered easily. This type of the probability statement will play an important role in making inferences about the unknown population mean (μ) later in Chapter 7.

In practice, the size of population (N) is often infinite or very large, and thus it is not even possible to list all possible samples of size n ($n < N$) as we did in Table 6.2. Consequently, deriving the sampling distribution for the sample mean is not feasible. Fortunately, there are known theories (rules) about the sampling distribution of $\bar{X}$ that we can use to find the associated probabilities, without computing the values of $\bar{X}$ based on all possible samples. There are two different cases:

1. When the population distribution is normal, i.e., when $X \sim N(\mu, \sigma^2)$.

2. When the sample size (n) is large

In case 1, when the distribution of the population of X is normal, the (sampling) distribution of $\bar{X}$ is is also normal regardless of the sample size. The mean and variance of the distribution of $\bar{X}$ are given in Equations 6.1 to 6.3. For normal sampling the sample correction is zero; thus, the finite correction factor in Equation 6.3 can be ignored and the distribution of $\bar{X}$ can be stated as follows:

$$\bar{X} \sim N(\mu, \sigma^2/n). \tag{6.4}$$

Equivalently, Equation 6.4 can be expressed as

$$\frac{\bar{X} - \mu}{\sigma/\sqrt{n}} \sim N(0,1). \tag{6.5}$$

This is an important result and forms the basis of many of the statistical tests used by transportation professionals. However, an astute reader will see a flaw in this logic: If we are trying to examine the population mean, how likely is it for us to know the population variance? The answer is that in most cases we will not know this. A natural solution would be to substitute the sample variance for the population variance. A considerable amount of work has been conducted over the past one hundred years to understand the implications of this substitution. If the population standard deviation (σ) is replaced by the sample standard deviation (S) in Equation 6.5, the resulting distribution will be more spread out than the standard normal distribution due to extra variability introduced by S. The sampling distribution of the standardized $\bar{X}$ in that case is given as the t-distribution with $n - 1$ degrees freedom:

$$\frac{\bar{X} - \mu}{S/\sqrt{n}} \sim t_{n-1}. \tag{6.6}$$

This distribution is sometimes referred to as a Student's t-distribution after its discoverer, William Sealy Gosset, who wrote under the name *Student*.

In case 2, there is a well-known theorem in statistics called the central limit theorem (CLT). This theorem describes the sampling distribution of $\bar{X}$ regardless regardless of the underlying population distribution as long as the population variance σ^2 exists (i.e., is finite). The theorem can be formally stated as follows:

CENTRAL LIMIT THEOREM

If $\bar{X}$ is the sample mean obtained from a random sample of size n of which the population distribution has a mean μ and variance σ^2, then the distribution of $\bar{X}$ will approach the normal distribution with mean μ and variance σ^2/n as n increases.

$$\bar{X} \sim N(\mu, \sigma^2/n).$$

Using the central limit theorem, we say that the sampling distribution of $\bar{X}$ can be approximated reasonably well by a normal distribution if n is large enough. A general rule of thumb is that "large enough" corresponds to n greater than 30 as long as the underlying population distribution is not too skewed. If the underlying population distribution is severely skewed, a number much larger than thirty would be needed to apply the CLT. It cannot be stressed enough that the CLT holds regardless of the shape of the population distribution (as long as the variance exists). The population distribution may be discrete or continuous, as well as symmetric or skewed. Nevertheless, we can make inferences about the mean of the population without depending on its distribution, and this is why the CLT is often considered the most important theorem in statistics. You will see variations of the above equations throughout the remainder of the text whenever inferences on population means are required.

The central limit theorem on the sample mean can be illustrated using the SA loop speed data (recorded in mph) given in Figure 3.1. The histogram of the original data shows that the speed data in this case are not bell shaped and are skewed left. Recall that the mean and variance of the original data were 49.059 mph and 158.253 mi^2/h^2, respectively.

To illustrate the CLT, assume that we take a random sample of size n from the data given in Figure 3.1 and calculate the sample mean. Then we repeat this one thousand times and construct a histogram. Figure 6.1 shows the results for a value of n of 5, 15 and 30.

As can be observed from Figure 6.1, the distribution of $\bar{X}$ becomes more bell shaped as n increases and approaches a normal distribution. As stated earlier, this is a general result that is independent of the underlying distribution. Also, notice that the histograms of $\bar{X}$ are all centered on the population mean 49.05. It is also apparent that the variability in the sampling distribution of $\bar{X}$ decreases as n increases. The values $\hat{\mu}_{\bar{X}}$ and $\hat{\sigma}_{\bar{X}}^2$ given under each histogram represent the sample mean and variance of those one thousand $\bar{X}$ values. As we discussed previously, the variance of the sampling distribution of $\bar{X}$ can be shown to be $\sigma_{\bar{X}}^2 = \sigma^2/n$ if it is computed based on infinitely many samples. Although, we have one thousand samples, which can be considered many but not infinite, the estimated variance of the sampling distribution of $\bar{X}$, $\hat{\sigma}_{\bar{X}}^2$ is still very close to its theoretical value $158.253/n$ for each of $n = 5$, 15, and 30.

When the population distribution is nonnormal and the sample size is small, we do not have any general theorem for describing the distribution of $\bar{X}$ because it depends on the population distribution and can be different for each population. In this situation you will not easily be able to make an assessment about the quality of your estimate. There may be some options, but you will need more statistical knowledge than provided in this textbook. Under these situations, the authors recommend that you seek the services of a professional statistician.

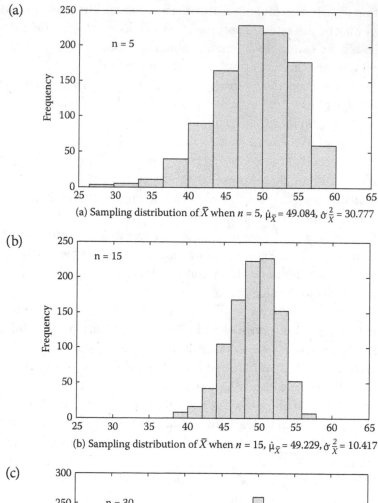

(a) Sampling distribution of $\bar{X}$ when $n = 5$, $\hat{\mu}_{\bar{X}} = 49.084$, $\hat{\sigma}^2_{\bar{X}} = 30.777$

(b) Sampling distribution of $\bar{X}$ when $n = 15$, $\hat{\mu}_{\bar{X}} = 49.229$, $\hat{\sigma}^2_{\bar{X}} = 10.417$

(c) Sampling distribution of $\bar{X}$ when $n = 30$, $\hat{\mu}_{\bar{X}} = 49.161$, $\hat{\sigma}^2_{\bar{X}} = 5.279$

FIGURE 6.1 Frequency histogram for $\bar{X}$ based on one thousand random samples.

6.4 SAMPLING DISTRIBUTION OF A SAMPLE VARIANCE

In Chapter 3, the sample variance s^2 was introduced as a measure of variability in the data. When the data are obtained as a random sample from the population and the population variance σ^2 is unknown, the sample variance s^2 can be used to estimate σ^2. Similar to the fact that the value of the sample mean differs from sample to sample, the value of sample variance will also differ for different samples. This variability leads to the sampling distribution for the estimator s^2 as a random variable:

$$s^2 = \frac{\sum_{i=1}^{n}\left(X_i - \bar{X}\right)^2}{n-1}.$$

The sampling distribution of S^2 is usually given by considering the following ratio:

$$\frac{(n-1)s^2}{\sigma^2} = \frac{\sum_{i=1}^{n}\left(X_i - \bar{X}\right)^2}{\sigma^2}.$$

It can be shown that the ratio given above has a chi-square distribution with $n - 1$ degrees of freedom if X_i's $(i = 1, \cdots, n)$ are a random sample selected from a normal population with mean μ and variance σ^2. Though it is not demonstrated in this textbook, readers interested in the proof can refer to Section 5.4 of the book by Casella and Berger (1990). This sampling distribution will be used as a basis for making inferences about the unknown population variances in Chapter 7. Note that if you do not know the underlying population distribution, then nonparametric methods—such as the bootstrap confidence intervals described in Chapter 13—will need to be used to make inferences about the sample variance. Interested readers may refer to the book by Keeping (1995).

6.5 SAMPLING DISTRIBUTION OF A SAMPLE PROPORTION

As discussed in Chapter 3, a sample proportion is a summary statistic for the categorical data, which can be viewed as the sample mean if there are only two categories labeled as 1 (success) and 0 (failure) in the data. We used a notation p earlier to represent a proportion of observations corresponding to a certain category in the data (a sample proportion unless the data are the population) in Chapter 3, and also to represent the probability of success in the binomial distribution (a population proportion) in Chapter 5. To avoid any confusion between the sample proportion and the population proportion, we are going to reserve p to denote the population proportion hereafter, and use $\hat{p}$ to denote the sample proportion. As in the case of the sample mean and the sample variance, the value of the sample proportion ($\hat{p}$) will be different from sample to sample, and this variability leads to the sampling distribution for the estimator $\hat{P}$ as a random variable.

Let us think of a population consisting of all drivers using toll roads, and define a random variable X representing the gender of a driver using toll roads as follows:

$$X_i = \begin{cases} 0 \text{ if a driver is a male} \\ 1 \text{ if a driver is a female.} \end{cases}$$

Let p denote the population proportion of female drivers using toll roads:

$$p = Prob(X = 1).$$

Suppose we select a random sample of size, $n, \{X_1, X_2, \cdots, X_n\}$, consisting of drivers using toll roads and count the number of 1s (female drivers) in the sample to estimate p. In Chapter 5, we learned that the number of successes in n independent trials can be modeled using a binomial distribution. If we consider that the observation of a female driver is a success, the number of female drivers in a random sample of size n, $\sum_{i=1}^{n} X_i$, will follow a binomial distribution. That is,

$$\sum_{i=1}^{n} X_i \sim B(n, p).$$

The sample proportion $\hat{p}$ obtained by dividing the number of female drivers in the sample by n takes a form of a sample mean as follows:

$$\hat{P} = \frac{\sum_{i=1}^{n} X_i}{n}.$$

Recall from Chapter 5 that the mean and variance of the binomial distribution $Bin(n, p)$ are np and $np(1-p)$, respectively. That is,

$$E\left(\sum_{i=1}^{n} X_i\right) = np$$

$$Var\left(\sum_{i=1}^{n} X_i\right) = np(1-p).$$

Using the properties of mean and variance of random variables discussed in Chapter 4, we can show that

$$E(\hat{P}) = E\left(\frac{1}{n}\sum_{i=1}^{n} X_i\right) = \frac{1}{n}E\left(\sum_{i=1}^{n} X_i\right) = \frac{1}{n} \times np = p \qquad (6.7)$$

and

$$Var(\hat{P}) = Var\left(\frac{1}{n}\sum_{i=1}^{n} X_i\right) = \left(\frac{1}{n}\right)^2 Var\left(\sum_{i=1}^{n} X_i\right) = \left(\frac{1}{n}\right)^2 \times np(1-p) = \frac{p(1-p)}{n}.$$

$$(6.8)$$

Similar to the distribution of the sample mean, the distribution of the sample proportion is centered at its corresponding population parameter (the population proportion, p), and the variability is smaller than that of the population. The exact distribution of the sample proportion can be obtained from a binomial distribution; however, common statistical software will be required to calculate the binomial probabilities. If this were being done by hand it would be a rather tedious exercise.

Because the sample proportion is also a sample mean, the central limit theorem (CLT) that we discussed earlier can be applied when n is sufficiently large and the distribution of the sample proportion can be approximated by a normal distribution. This is called the normal approximation to the binomial distribution. The rule of thumb for determining a large enough n for the CLT to be applied, however, is not $n > 30$ anymore. How large n needs to be depends on p because the shape of underlying population distribution varies considerably with p. For a value of p that is close to 0 or 1, the population distribution is severely skewed and requires a much larger sample to apply the CLT than that for p close to 0.5, for which the population distribution is not severely skewed. A commonly used rule of thumb for approximating the distribution of the sample proportion by a normal distribution is $np \geq 5$ and $n(1-p) \geq 5$. This rule, however, may not work well for some combinations of n and p. See Brown et al. (2001) for more discussions on this topic. If p is unknown, then $\hat{p}$ may substitute for p.

We can summarize the sampling distribution of $\hat{p}$ as follows:

SAMPLING DISTRIBUTION OF $\hat{p}$

The mean $\mu_{\hat{p}}$ and variance $\sigma_{\hat{p}}^2$ of a sample proportion $\hat{p}$ obtained from a random sample of size n from a population having the population proportion, p, are

$$\mu_{\hat{p}} = E(\hat{P}) = p$$

and

$$\sigma_{\hat{p}}^2 = Var(\hat{P}) = \frac{p(1-p)}{n}.$$

In addition, if $n\hat{p} \geq 5$ and $n(1-\hat{p}) \geq 5$, the distribution of $\hat{p}$ may be approximated by a normal distribution with mean $\mu_{\hat{p}}$ and variance $\sigma_{\hat{p}}^2$, i.e., $\hat{p} \sim N(p, p(1-p)/n)$.

6.6 CONCLUDING REMARKS

Sampling is a difficult concept to learn. This chapter is intended to explain the difference between samples and populations. Sample statistics, such as the sample mean and sample variance, change from sample to sample. Their changes follow the sampling distributions explained in this chapter. Readers need to understand this so that inferences made from samples in the next chapters are sensible. Sampling distributions will be the skeleton that supports the body of our inferential methods.

Transportation professionals are often asked specific questions about the operations of the transportation system, such as:

Are drivers speeding on a particular roadway?

What is the percentage of high-income drivers who use a high-occupancy toll lane?

Is the highway system operating under HCM level of service C?

If the engineer had complete knowledge of the system, then these types of questions would be readily answered. Unfortunately, this is not often the case, and the engineers have to obtain samples to answer the questions. Most transportation professionals are familiar with the concepts of sample mean and sample variance. Estimates derived from the sample can be used to answer these types of questions.

A more subtle question, however, is: How good is the estimate in answering the question? If you asked someone to analyze whether drivers were speeding and she gave you a sample mean speed based on two observations, you might feel that the estimate would not be appropriate. However, just how many observations are required is still an open question.

To fully answer the underlying question, you essentially need to know the sampling distribution. As was introduced in this chapter, the sample mean and sample variance are both random variables that have a distribution (sampling distribution). It was shown that under certain circumstances we will know (or can approximate) this distribution. Once we have this distribution, we can then make inferences about how good our answer is—this will form the basis of the following chapters. It cannot be emphasized enough that these techniques can only be used under certain circumstances. As a professional, you will be responsible for understanding when they can be used.

In summary, this chapter will form the basis of most of the work in the following chapters; therefore, the reader is cautioned to understand these concepts fully before proceeding. Alexander Pope's famous quote "A little learning is a dangerous thing" applies particularly well to statistics. While statistical formulas may be readily used, particularly with today's software, such as JMP, it is imperative that users have a full understanding of the underlying assumptions lest they get themselves, or their clients, in erroneous situations.

HOMEWORK PROBLEMS

1. What is the probability that the mean ($\bar{X}$) of a random sample of size 2 selected from the population in Table 6.1 lies within 3 mph of the population mean ($\mu = 61.6$)?

2. Suppose an engineer wants to find the percentage of commercial vehicles on a roadway segment. In the first hour there were forty-seven vehicles, and five of them were commercial vehicles.

   a. Based upon the first forty-seven vehicles, what percent are commercial?

b. Using a binomial distribution, what is the chance that the population percentage is greater than 15%?

c. Using a normal approximation to the binomial distribution, what is the chance that the population percentage is greater than 15%?

d. Explain the differences in your answers for b and c.

3. Probe vehicles are used to monitor speeds of several freeway segments. Describe deviations from random sampling for each of the following:

a. Probe-vehicle drivers are to stay in the middle of the pack of vehicles with which they enter the freeway.

b. Probe-vehicle drivers are to pick out a car of their choice and follow it at a safe distance.

c. Probe vehicles are dispatched at a random time, to an intersection chosen randomly, and a computer randomly chooses which vehicle they are to follow at a safe distance.

4. The average fuel efficiency for a vehicle driving on a highway is 21 mpg with a standard deviation of 4 mpg. What is the probability that while driving a 10-mile segment of highway, one hundred randomly chosen vehicles use a total of more than 58 gallons of fuel? (Hint: A sum is $n \times \bar{X}$.)

5. A state DOT wants to audit contracts to assess the percent that have had all requirements fulfilled within the last five years.

a. What is the population of interest?

b. How should the contracts be chosen so that the audited contracts can be used to make inferences about the population?

c. How many contracts need to be audited to attain a standard deviation for the estimated percentage of compliant contracts less than or equal to .05? (Hint: The standard deviation of $\hat{p}$ is $\sqrt{p(1-p)/n}$. The value of p that gives the largest standard deviation is .5.)

REFERENCES

Brown, L. D., T. T. Cai, and A. DasGupta. 2001. Interval estimation for a binomial proportion. *Statistical Science* 16:101–33.

Casella, G., and R. L. Berger. 1990. *Statistical inference.* Pacific Grove, CA: Wadsworth Brooks/Cole.

Cochran, W. G. 1977. *Sampling techniques.* 3rd ed. New York: John Wiley & Sons.

Keeping, E. S. 1995. *Introduction to statistical inference.* N. Chelmsford, MA: Courier Dover Publications.

Lohr, S. L. 1999. *Sampling: Design and analysis.* Pacific Grove, CA: Duxbury Press.

CHAPTER 7

Inferences

Hypothesis Testing and Interval Estimation

7.1 INTRODUCTION

Often transportation professionals are asked to make specific statements regarding the transportation system. For example:

Are average speeds on a freeway within federal guidelines?

Has a safety improvement actually led to a reduction in crashes?

Are the contractors meeting the specifications in the contract?

Are ambulances reaching their destinations within their target time?

Is congestion on one roadway worse than another?

Over the years much effort has been exerted to build statistical models that help answer these types of questions. Curve fitting and density estimation (see appendix) are examples of model building. A branch of statistics that is dedicated to testing whether models are consistent with available data will be the focus of this chapter. Generally, this branch goes under the twin headings of hypothesis testing (that will be introduced in this chapter) and goodness of fit (which will be covered in Chapter 9). Excellent books have been written on the subject; see, for example, the classic text by Lehmann and Romano (2005).

Many contemporary statisticians believe and say "no model is exact but some models are useful." The authors believe the last statement is correct and that the value in testing comes from measuring the concordance of the model with data. Complete concordance should not be expected. Nonetheless, we first present classical hypothesis testing as it underlies the construction of confidence intervals (CIs). These tests basically provide yes or no answers to the types of questions listed above. We then introduce the concept of confidence intervals that provide a range of plausible models. However, it should be noted that while both hypothesis tests and CI will give identical answers to the same questions, the authors believe confidence intervals are more informative. Hence, confidence intervals are the recommended approach.

7.2 FUNDAMENTALS OF HYPOTHESIS TESTING

Our approach to hypothesis testing starts by partitioning all models into two sets. Let θ be either a model or an alias for a model. For example, θ can be a mean, a regression function, or a probability distribution. There are two competing hypotheses in hypothesis testing: the null and the alternative hypotheses. An alternative, or research, hypothesis (denoted by H_a) is what we would like to prove using the data. The null hypothesis (denoted by H_0) is negation of the research hypothesis. Let the set of all models under consideration be Θ. We start hypothesis testing by assuming that Θ is the union of disjoint sets:

$$\Theta_{H_0} \quad \text{and} \quad \Theta_{H_a}.$$

That is,

$$\Theta_{H_0} \cup \Theta_{H_a} = \Theta \quad \text{and} \quad \Theta_{H_0} \cap \Theta_{H_a} = \varnothing, \text{ the empty set.}$$

The models represented by Θ_{H_0} usually are models that are initially assumed to be true, but we want to prove are inappropriate.

Based on the amount of evidence in the data (as discussed in the next section), we may or may not reject the null hypothesis. Rejecting H_0 or not rejecting H_0 are two possible decisions that can be made by a researcher and could be correct or incorrect depending on the true nature of the variable. Table 7.1 shows that there are two types of errors encountered in hypothesis testing: type I and type II errors.

Type I error is the error of rejecting H_0 when H_0 is true.

Type II error is the error of not rejecting H_0 when H_0 is false.

TABLE 7.1   Outcomes of Hypothesis Testing

		True Nature	
		H_0 Is True	H_0 Is False
Decision	Reject H_0	Type I error	Correct
	Fail to reject H_0	Correct	Type II error

The probability of a type I error (i.e., the probability of rejecting H_0 when it is true) is denoted by α, and the probability of a type II error (i.e., the probability of not rejecting H_0 when it is false) is denoted by β. The probability of correctly rejecting H_0 (i.e., the probability of rejecting H_0 when it is false) is $1 - \beta$ and is called the power of the test. We wish to obtain the test that minimizes both α and β. It is, however, not possible to minimize both error rates in hypothesis testing for a fixed sample size because there is a trade-off between α and β (the smaller α is, the larger β is, and vice versa). The only way of decreasing both error rates is to increase the sample size or decrease the measurement variance. A test that never rejects H_0 has a type I error rate of 0, but a type II error rate of 1. On the other hand, a test that always rejects H_0 has a type II error rate of 0, but a type I error rate of 1. To get around this problem, we usually specify the type I error rate as the maximum value that we can tolerate (for example, $\alpha = 0.05$ or 0.1), and then obtain the corresponding test.

Suppose that we are told that a particular highway section has a speed limit of 65 mph, and that the average speed of vehicles is equal to this speed limit. Based on our experience, we may suspect that an average speed is not 65 mph. In this situation our alternative hypothesis is that the average speed is not 65 mph, or $\Theta_{H_a} = \{\mu | \mu \neq 65\}$ (this is, the set of all average speeds that are not equal to 65 mph). The null hypothesis is that the average speed is 65 mph ($\mu = 65$), and we initially assume that the null hypothesis is true. In statistical hypothesis terms, we have the null hypothesis $H_0 : \mu = 65$ mph vs. the alternative hypothesis $H_a : \mu \neq 65$ mph.

It is instructive to pursue this average speed example. First, we must define what we mean by average speed. Do we mean that the weekly average, monthly average, weekend average, or today's average is 65 mph? This is a nontrivial question. Do we believe that the average is stable or stationary over time? If it is not, then how can we hope to gather data to test it? Moreover, how is speed to be measured and how biased are the measurements? For example, are double-loop inductance detectors and radar guns both used to measure speeds? Are both measurement methods subject to the same biases? How big is the random error in the speed measurements, and is the standard deviation of a speed measurement related to the true speed? Are average speeds subject to time trends; for example, are speeds slower during rush hour or bad weather?

These are examples of important questions that transportation professionals must be cognizant of when using standard statistical hypothesis testing tools. Most elementary statistic books treat generic cases that are easy—that is, where data are assumed to fit unrealistically simple models. We will go a bit deeper into these issues in this chapter because even relatively mundane transportation questions can be quite complex.

7.3 INFERENCES ON A SINGLE POPULATION MEAN

7.3.1 Hypothesis Tests about a Population Mean

We start our technical exposition by assuming we have n independent measurements of the continuous variable speed, $X_1, X_2, \ldots X_n$, that have a random error with variance σ^2. The model is

$$X_i = \mu + \sigma \varepsilon_i.$$

The parameter μ represents the population mean of the measurements, and in this case, this would be the true value of average speed. For the sake of simplicity, we will assume the errors ε_i, $i = 1, \ldots, n$, are normally distributed with mean 0 and variance 1. This is a standard model given in most introductory statistics courses. It does not include many of the features discussed in the previous section; for example, there is no bias accommodated by this model. The null hypothesis is denoted as H_0 is $H_0 : \mu = \mu_0$, where μ_0 is a given value (for instance, at 65 mph). The alternative hypothesis is denoted by H_a is $H_a : \mu \neq \mu_0$. Note that we focus first on the null hypothesis of equality and the alternative hypothesis of nonequality. We then discuss one-sided hypotheses such as

$$H_0 : \mu \leq \mu_0 \text{ vs. } H_a : \mu > \mu_0$$

or

$$H_0 : \mu \geq \mu_0 \text{ vs. } H_a : \mu < \mu_0.$$

Typical statistical approaches estimate the mean parameter, μ, by a statistic $\bar{X}$ and choose the more reasonable of the two competing hypotheses based upon how close $\bar{X}$ is to μ_0. Clearly, H_0 is rejected if μ_0 and $\bar{X}$ are too far apart. In this chapter the measure of closeness depends upon the anticipated variability of $\bar{X}$ (i.e., standard error of $\bar{X}$). For example, it is impossible to say if a difference of 5 mph is close or far without having an understanding of the variance of the statistic $\bar{X}$.

We learned in Chapter 6 that the standard error of $\bar{X}$ is $\sigma/\sqrt{n}$. Thus, if σ is known, the typical measure of distance between $\bar{X}$ and μ_0 will be

$$d_\sigma(\mu_0, \bar{X}) = \frac{|\bar{X} - \mu_0|}{\sigma/\sqrt{n}}.$$

Recall that the distribution of $\bar{X}$ is normal either when X follows a normal distribution or when n is large enough. Let $Z = (\bar{X} - \mu_0)/(\sigma/\sqrt{n})$. Then under the null hypothesis (i.e., assuming $\mu = \mu_0$), we have by Equation 6.5, in which

$$Z = \frac{(\bar{X} - \mu_0)}{\sigma/\sqrt{n}} \sim N(0,1)$$

if either normality or a large sample size condition is satisfied. Let $z_{\alpha/2}$ and $z_{1-\alpha/2}$ be the lower $\alpha/2$ percentile and the upper $\alpha/2$ percentile of the standard normal distribution, respectively. Then, from the property of the standard normal distribution, we know that

$$P(z_{\alpha/2} \le Z \le z_{1-\alpha/2}) = P(-z_{1-\alpha/2} \le Z \le z_{1-\alpha/2}) = P(|Z| \le z_{1-\alpha/2}) = 1 - \alpha. \quad (7.1)$$

If $\alpha = 0.05$, Equation 7.1 becomes

$$P(z_{.025} \le Z \le z_{0.975}) = P(|Z| \le 1.96) = P\left(\frac{|\bar{X} - \mu_0|}{\sigma/\sqrt{n}} \le 1.96\right) = 0.95. \quad (7.2)$$

Equation 7.2 implies that if H_0 is true most of the time (i.e., 95% of the time), the standardized distance of $\bar{X}$ from μ_0, $d_\sigma(\mu_0, \bar{X}) = |\bar{X} - \mu_0|/(\sigma/\sqrt{n})$, is at most 1.96 ($z_{0.975}$). If the observed distance $d_\sigma(\mu_0, \bar{X}_{observed})$ is greater than 1.96, then it may be considered a rare event (less than 5% of chance) under the null hypothesis. Thus, we may reject H_0, although there is a small chance (5% here) of incorrectly rejecting H_0. This defines a hypothesis test about the population mean, μ, with a type I error probability $\alpha = 0.05$.

Transportation students often are confused about the exact interpretation of these tests. What do statisticians mean when they say that the null hypothesis is rejected at $\alpha = 0.05$. They mean that when the null hypothesis is true, if one hundred random samples of the same size were taken, then, on average, the hypothesis test would provide the correct result ninety-five times. (That is, they would 95% of the time not reject the null hypothesis.) Because we are only taking one sample, we can only make probabilistic statements about our results. For example, if the true mean speed were 65 mph, and that were also the null hypothesis and we repeated the experiment one hundred times, we would expect to incorrectly reject the null hypothesis (on average) five times.

7.3.1.1 z-Test

Oftentimes $Z = (\bar{X} - \mu_0)/(\sigma/\sqrt{n})$ is referred to as a test statistic. The hypothesis test based on $Z = (\bar{X} - \mu_0)/(\sigma/\sqrt{n})$ is called a z-test. Depending on the form of the alternative hypothesis, there could be three different z-tests with a type I error probability α:

1. When the alternative hypothesis is H_a: $\mu \neq \mu_0$, the z-test rejects H_0 if and only if

$$\frac{|\bar{X} - \mu_0|}{\sigma/\sqrt{n}} \geq z_{1-\alpha/2}.$$

2. When the alternative hypothesis is H_a: $\mu > \mu_0$, we reject the null hypothesis when

$$\frac{(\bar{X} - \mu_0)}{\sigma/\sqrt{n}} \geq z_{1-\alpha}.$$

3. When the alternative hypothesis is H_a: $\mu < \mu_0$, we reject the null hypothesis when

$$\frac{(\bar{X} - \mu_0)}{\sigma/\sqrt{n}} \leq z_{\alpha}.$$

Cases 2 and 3 would be used if we were interested in whether some metric (e.g., average speed) were greater or lesser than some critical boundary (e.g., the speed limit). These types of questions—Are people speeding on average? Is the pavement thickness less than specified?—are often encountered in transportation engineering statistics. Case 1 would be typical of questions where we wonder if the metric is higher or lower than some specified criteria, such as in instances of air voids in concrete.

7.3.1.2 t-Test

In most transportation applications σ is not known. In this case the procedure is essentially the same. The sample standard deviation (S) replaces the population standard deviation, σ, and a t-test is used in place of the z-test. This follows from the discussion in Chapter 6 in which the distribution of the sample mean when the variance is unknown follows a t-distribution. Recall from Equation 6.6 that

$$\frac{\bar{X} - \mu}{S/\sqrt{n}} \sim t_{n-1}.$$

Thus, a two-sided t-test rejects H_0 if and only if $d_S(\mu_0, \bar{X}) > t_{n-1,1-\alpha/2}$, where $t_{n-1,1-\alpha/2}$ is the upper $\alpha/2$ percentile from a t-distribution with $n - 1$ degrees of freedom. One-sided tests are handled as above. When the alternative hypothesis is $H_a : \mu > \mu_0$, we reject the null hypothesis when

$$\frac{(\bar{X} - \mu_0)}{S/\sqrt{n}} \geq t_{n-1,1-\alpha},$$

and when the alternative hypothesis is $H_a : \mu < \mu_0$, we reject the null hypothesis when

$$\frac{(\bar{X} - \mu_0)}{S/\sqrt{n}} \leq t_{n-1,\alpha}.$$

For z- and t-tests, we have the summary shown in Table 7.2 for the tests used.

Sometimes p values rather than percentiles of normal or t-distributions are used in hypothesis testing. The p value for these tests is given by

$$p \text{ value} = P\left(d_\sigma(\mu_0, \bar{X}) \geq d_\sigma(\mu_0, \bar{X}_{observed})\right)$$

or

$$p \text{ value} = P\left(d_S(\mu_0, \bar{X}) \geq d_S(\mu_0, \bar{X}_{observed})\right),$$

respectively. As can be seen from above, assuming H_0 is true, the p value can be interpreted as the probability of obtaining a relative distance (or a test statistic value) as extreme as or more extreme than the relative distance between the hypothesized mean and the observed sample mean (or the observed test statistic value).

TABLE 7.2 Summary of Tests for One Population Mean

		Hypotheses	Rejection Regions for the Tests
z-Test	Case 1 (σ: known)	$H_0 : \mu \leq \mu_0$ vs. $H_a : \mu > \mu_0$	Reject H_0 if $\dfrac{(\bar{X} - \mu_0)}{\sigma/\sqrt{n}} \geq z_{1-\alpha}$
	Case 2 (σ: known)	$H_0 : \mu \geq \mu_0$ vs. $H_a : \mu < \mu_0$	Reject H_0 if $\dfrac{(\bar{X} - \mu_0)}{\sigma/\sqrt{n}} \leq z_{\alpha}$
	Case 3 (σ: known)	$H_0 : \mu = \mu_0$ vs. $H_a : \mu \neq \mu_0$	Reject H_0 if $\dfrac{|\bar{X} - \mu_0|}{\sigma/\sqrt{n}} \geq z_{1-\alpha/2}$
t-Test	Case 4 (σ: unknown)	$H_0 : \mu \leq \mu_0$ vs. $H_a : \mu > \mu_0$	Reject H_0 if $\dfrac{(\bar{X} - \mu_0)}{S/\sqrt{n}} \geq t_{n-1,1-\alpha}$
	Case 5 (σ: unknown)	$H_0 : \mu \geq \mu_0$ vs. $H_a : \mu < \mu_0$	Reject H_0 if $\dfrac{(\bar{X} - \mu_0)}{S/\sqrt{n}} \leq t_{n-1,\alpha}$
	Case 6 (σ: unknown)	$H_0 : \mu = \mu_0$ vs. $H_a : \mu \neq \mu_0$	Reject H_0 if $\dfrac{|\bar{X} - \mu_0|}{S/\sqrt{n}} \geq t_{n-1,1-\alpha/2}$

A common error is to interpret the *p* value as the probability that the null hypothesis is true. Because the *p* value depends upon the sample size, this is incorrect. If the *p* value for a test is less than or equal to α, then the null hypothesis is rejected. The *p* value provides a more rational description of the weight of the evidence than does a simple reject or do not reject statement. Consider a test where the *p* value is 4.9%, and another test in which the *p* value is 5.1%. In the former, the null hypothesis would be rejected. In the latter, however, it would not be rejected. From an engineering perspective, the analyst might decide that these values are close enough and further investigation may be warranted. A strict yes or no answer using hypothesis testing would not provide this insight.

These tests and discussion of *p* values are provided in most introductory statistics books, such as in Ang and Tang (2006) or Ott and Longnecker (2009). Note that when sample sizes are very large, differences in population can be detected even if those disparities are unimportant to engineers. This is a critical point since many engineers will experience this in practice. That is, it is easy to show statistical differences between the sample mean and a critical value by taking a large number of samples, even though this is not of critical interest to the engineer. For example, if the true population mean is 65.5 mph on the roadway and the speed, or speed limit, you are interested in is 65 mph, the engineer may argue there is no practical difference between the speeds: the cars are following the speed limit on average. However, with a sample size around 400 and a measurement standard deviation of about 3 mph, a statistical difference could be shown between the observed speeds and the speed limit. While statistically significant, this difference is of no practical importance. There are two dangers to be careful of in practice. The first is to keep taking observations until there is a statistical difference. That is, you did not find a statistically significant difference when one hundred observations were taken. Consequently, you took nine hundred more and found a statistical difference. The second danger occurs when someone (e.g., a contractor) needs to show a statistically significant difference according to some pre-scribed quality control criterion. In this situation they may continue to test until there is a statistically significant difference. Unfortunately, the theory needed to handle sequential testing is well beyond the scope of this book. By using common sense and a good knowledge of the underlying assumptions these issues can be avoided. Oftentimes the number of samples is selected *a priori*, which effectively removes the temptation to keep taking samples until a statistically significant result is identified.

In summary, the drawbacks to the hypothesis tests include:

1. They give only a yes or no answer to the question of whether H_0 should be rejected. They do not provide information on what a reasonable range of values is for μ.

2. Because most people do not believe such a simple null hypothesis could possibly be true (for example, could $\mu = 65$ mph to a million decimal places?), the only information coming from the test is information about power. That is, we find out if the test made the correct decision (one that could have been made without any data); the null hypothesis should be rejected. The power of a test is determined by the sample size and measurement precision. How valuable is it to know that a sample size was too small to reject a hypothesis that is known to be false?

7.3.2 Interval Estimates for a Population Mean

Instead of testing a hypothesis, it is frequently more useful to estimate a parameter and specify a plausible range of values for the parameters. Typically, the plausible range is given as a confidence interval. While confidence intervals are closely related to hypothesis tests, they are used to approach the issue from a different point of view. Furthermore, confidence intervals are typically more easily modified to handle bias. To illustrate these issues, we consider confidence intervals for average speed.

If we know the value of σ, then the typical $(1-\alpha)100\%$ confidence interval is given as

$$\bar{X} \pm z_{1-\alpha/2}\sigma/\sqrt{n}.$$

If we do not know σ, then the $(1-\alpha)100\%$ confidence interval is given as

$$\bar{X} \pm t_{n-1,1-\alpha/2}S/\sqrt{n}.$$

Note that the latter case is more typically found in transportation engineering. Intuitively, if we knew the population variance, we probably would know the population mean as well, which would remove the need for our hypotheses testing.

Once the confidence intervals are calculated, they define the results of an infinite set of hypothesis tests *a priori*. Basically, H_0 is rejected whenever μ_0 is not in the confidence interval. By definition, t-tests (or z-tests) only give simple "reject" or "do not reject" answers. It cannot be emphasized enough that, if done correctly, the hypothesis tests and the confidence intervals will give the same answer to the hypothetical question we are examining. However, as will be demonstrated in this chapter, the latter give much more information and provide a more intuitive feel for the data.

7.3.2.1 Bias-Adjusted Confidence Intervals

It is easy to modify the confidence interval to accommodate bias. Suppose that the speed measuring device may be biased. For example, we may know that the device is only accurate to within $\pm B$ mph. That is, $EX = \mu + b$, and the bias b is known to satisfy the bound $|b| \leq B$. Then, the corresponding bias adjusted confidence intervals are

$$\bar{X} \pm \left(B + z_{1-\alpha/2}\sigma/\sqrt{n}\right)$$

and

$$\bar{X} \pm \left(B + t_{n-1,\alpha/2}S/\sqrt{n}\right)$$

respectively. This adjustment for bias is conservative; that is, the intervals are too wide. See Knafl et al. (1984) for a more accurate and complicated approach. The impact of adjusting the confidence intervals for bias makes the intervals more realistic. As opposed to traditional confidence intervals, the formula also shows that the width of the intervals decreases more slowly as the sample size increases. Accordingly, the bias-adjusted intervals never decrease to zero. For many transportation applications this is a more realistic scenario. For example, many measuring devices have a known bias that can be readily obtained from the manufacturer or device manual. Large sample sizes in the presence of bias do not guarantee estimation accuracy. For a CI that ignores the effect of bias, large sample sizes might give a false sense of the accuracy of the analysis. Importantly, this cannot happen when bias is correctly taken into account.

Example 7.1

To illustrate these methods without a bias adjustment we consider speed data in Houston. The data are given in Table 7.3. Consider that we are interested in testing whether the true mean speed is 65 mph or not. The speed data are assumed to follow a normal distribution.

Using t-tests (because the population standard deviation is unknown) to test the hypothesis that H_0: $\mu = 65$ vs. H_a: $\mu \neq 65$ in JMP supplies the results shown in Figure 7.1.

The average speed observed in the field is 64.24 mph and the hypothesized value is 65 mph. The question is whether the value of 64.24 is far enough away from 65. Based on the t-test, it is clear from the null hypothesis that the mean speed of 65 mph cannot be rejected.

The corresponding 95% confidence interval that is not corrected for bias is given in Figure 7.2.

TABLE 7.3 Speed Data in Houston for Testing for the Mean Speed

Observation No.	Speed
1	64.636
2	64.344
3	51.522
4	65.229
5	54.903
6	67.075
7	61.558
8	84.643

▼ ☑ **Test Mean=value**

Hypothesized Value	65
Actual Estimate	64.2387
df	7
Std Dev	9.86759

t Test

Test Statistic	−0.2182
Prob > |t|	0.8335
Prob > t	0.5833
Prob < t	0.4167

FIGURE 7.1   Test statistics for testing mean speed.

N   8

▼ **Confidence Intervals**

Parameter	Estimate	Lower CI	Upper CI	1−Alpha
Mean	64.23875	55.98924	72.48826	0.950
Std Dev	9.867592	6.524197	20.08323	

FIGURE 7.2   Confidence intervals for mean speed and standard deviation of speed.

As expected, the CI gives the same result because the hypothesized value (65 mph) is within the lower (55.99 mph) and upper (72.49 mph) confidence limits. The power of the CI can easily be seen. If you also wanted to test whether the mean speed was 70 mph, then this can be done immediately with the CI by checking that 70 mph is also in the confidence interval. With t-tests the analysis would have to be recalculated.

Finally, if we knew that bias in speed measurements was no more than 3 mph, then we would expand both sides of the confidence interval by 3 mph. For example, if the hypothesized speed were 75 mph, we would reject this null hypothesis because it is not in the confidence interval we selected. However, if we included the bias term, the CI would be (52.99, 75.49) and we could not reject the null hypothesis that the average speed was 75 mph. Accordingly, by including bias we are less likely to reject a given null hypothesis, and this represents the uncertainty caused by the bias in the measuring device.

In most transportation applications, the assumption that the data are unbiased is hard to justify. Most speed sensors have some bias associated with them, and if this bias is known, then the CI can be adjusted to account for this. For example, it may be known that based on past experience the inductance loop detectors have up to a 2 mph bias when estimating speeds. Thus, the upper and lower bounds of the confidence intervals should be expanded to account for this bias. Hypothesis tests can also be directly adjusted for bias, but that is not pursued here. It is the authors' recommendation that you should use a bias-adjusted confidence interval in these situations.

7.4 INFERENCES ABOUT TWO POPULATION MEANS

The next most common question faced by engineers is whether two treatments or strategies are equivalent. For example, a transportation engineer might want to know whether the paint from two different manufacturers gives signs the same visibility. The engineer would first take a series of sign samples developed from each manufacturer and test them under various visibility conditions, such as rain and clear weather. It would be highly unlikely that the mean visibility (measured in feet from the point where the driver recognizes the sign) from manufacturer 1 would equal the mean visibility from manufacturer 2. The real question is how much different these two means would have to be before the engineer would feel comfortable stating they were dissimilar. This type of question often occurs in transportation engineering.

7.4.1 Hypothesis Tests about Equality of Two Population Means

We will deal with the case of testing equivalence of two means before handling the more general case of testing the equivalence of g-means with $g \geq 2$. The preceding discussion about the relationship among hypothesis tests, confidence intervals, and biases applies to testing equivalence of two means. Initially we will discuss the three cases that are usually handled by introductory statistics courses: the paired t-test, the pooled t-test, and the unequal variance t-test.

Suppose that we have measurement data from two populations. The measurements from the first are denoted by $X_1, \ldots, X_{n_1}$, and the data from the second population are denoted by $Y_1, \ldots, Y_{n_2}$. We denote the population means as μ_1 and μ_2, respectively. The hypotheses that we next consider are $H_0 : \mu_1 = \mu_2$ vs. $H_a : \mu_1 \neq \mu_2$. Just as in the single mean case, there are one-sided hypotheses that are of interest as well. Examples are $H_0 : \mu_1 \geq \mu_2$ vs. $H_a : \mu_1 < \mu_2$. A summary table is provided (Table 7.4) and includes the appropriate test for the most common hypotheses.

7.4.1.1 Paired t-Test

The paired t-test is designed to handle correlation among matched pairs of measurements or data points. For example, before and after measurements on the effects of alcohol upon driving performance (using the same drivers) would be tested and assessed using paired t-tests. In this case, the test subject is the same and only the presence of alcohol varies. A good indicator of the need for a paired t-test is to plot the paired data on a scatter plot. If there appears to be a linear trend in the data, then the paired t-test is the best choice among the three t-tests that we consider. However,

TABLE 7.4 Summary Table for T-Tests for Testing Equality of Two Population Means

Case	X-Y Pairs Use the Same Subject or Are Dependent Rejection Region	X-Y Pairs Are Independent and Have (Nearly) Equal Variances Rejection Region	X-Y Pairs Are Independent and Do Not Have (Nearly) Equal Variances Rejection Region
Case 1	$H_0: \mu_1 \leq \mu_2$ vs. $H_a: \mu_1 > \mu_2$	$t_{paired} \geq t_{n-1,1-\alpha}$	$t_{pooled} \geq t_{n_1+n_2-2,1-\alpha}$	$t_{\text{unequal variances}} \geq t_{df^*,1-\alpha}$
Case 2	$H_0: \mu_1 \geq \mu_2$ vs. $H_a: \mu_1 < \mu_2$	$t_{paired} \leq t_{n-1,\alpha}$	$t_{pooled} \leq t_{n_1+n_2-2,\alpha}$	$t_{\text{unequal variances}} \leq t_{df^*,\alpha}$
Case 3	$H_0: \mu_1 = \mu_2$ vs. $H_a: \mu_1 \neq \mu_2$	$\left|t_{paired}\right| \geq t_{n-1,1-\alpha/2}$	$\left|t_{pooled}\right| \geq t_{n_1+n_2-2,\alpha/2}$	$\left|t_{\text{unequal variances}}\right| \geq t_{n_1+n_2-2,\alpha/2}$

the requirement that the subjects are the same for paired observations is fairly rare in transportation applications. Therefore, the other two tests are much more common in transportation engineering.

7.4.1.2 Pooled t-Test

The pooled t-test is used to test the equivalence of two means when the data from the two populations are independent and have the same variance. For example, we may want to test the equivalence of the speeds approaching two different signalized intersections that are distant enough from each other as to be independent. We may also expect the luminescence readings of two different sign coatings to be independent. If the measurements have equivalent variances, then a pooled t-test is appropriate.

7.4.1.3 Unequal Variance t-Test

The unequal variance t-test is used to test the equivalence of two means when the data from the two populations are independent and may not have the same variance. As a practical matter, rarely are variances expected to be equal, but if the ratio of the two variances is far from 1, then an unequal variance t-test is appropriate. Later, we give tests for evaluating hypotheses about equal variances, but for now we will assume that the transportation professional knows *a priori* whether the variance of the measurements for the two populations is near equal or not.

All three t-tests that we use for two-sided hypotheses tests have the form

$$\frac{|\bar{X} - \bar{Y}|}{SE(\bar{X} - \bar{Y})},$$

where $SE(\bar{X}-\bar{Y})$ stands for the estimated standard error of the difference in the sample means. The three tests use different formulas to compute the estimated standard error, and the degrees of freedom associated with that estimated standard error.

The paired t-test assumes that the X-Y pairs of data are dependent, whereas the matched pairs (X_i, Y_i) for $i = 1,\ldots,n$ are independent as the subscript i changes. The paired t-test assumes that the data have a natural pairing and that $n_1 = n_2 = n$. Let $d_i = X_i - Y_i$. The standard error for the paired t-test is computed as

$$\sqrt{\frac{1}{n}} \sqrt{\sum_{i=1}^{n} (d_i - \bar{d})^2 / (n-1)},$$

and it has $n - 1$ degrees of freedom.

The pooled t-test assumes that all X and Y measurements are mutually independent, which also implies that all the X measurements and Y measurements are independent of each other. The standard error for the pooled t-test is computed using the formula

$$\sqrt{\frac{(n_1-1)S_X^2+(n_2-1)S_Y^2}{n_1+n_2-2}}\sqrt{\frac{1}{n_1}+\frac{1}{n_2}},$$

and it has $n_1 + n_2 - 2$ degrees of freedom.

The standard error for the unequal variance t-test is computed using the formula

$$\sqrt{\frac{S_X^2}{n_1}+\frac{S_Y^2}{n_2}}.$$

Its degrees of freedom involve a formula given in the appendix to this chapter. Fortunately, major statistics packages calculate the degrees of freedom calculation automatically.

Table 7.4 summarizes hypotheses and appropriate t-tests. The three t-statistics that we use are as follows:

$$t_{paired}=\frac{\sqrt{n}(\bar{X}-\bar{Y})}{\sqrt{\sum_{i=1}^{n}\left((X_i-Y_i)-(\bar{X}-\bar{Y})\right)^2/(n-1)}}$$

$$t_{pooled}=\frac{(\bar{X}-\bar{Y})}{\left(\sqrt{\frac{(n_1-1)S_X^2+(n_2-1)S_Y^2}{n_1+n_2-2}}\right)\sqrt{\frac{1}{n_1}+\frac{1}{n_2}}}$$

and

$$t_{unequal\ variances}=\frac{(\bar{X}-\bar{Y})}{\sqrt{\frac{S_X^2}{n_1}+\frac{S_Y^2}{n_2}}}.$$

Our t-test rejection regions are summarized in Table 7.4. The degrees of freedom for these tests are, respectively, $n-1$, n_1+n_2-2, and df^*. The formula for df^* is given in the appendix to this chapter.

7.4.2 Interval Estimates and Bias Adjustment for Difference of Two Population Means

The confidence interval that corresponds to the above three tests uses the formulas

$$\bar{X}-\bar{Y}\pm t_{df,1-\alpha/2}SE\left(\bar{X}-\bar{Y}\right).$$

The degrees of freedom used are the same ones used for the corresponding t-test. Just as in the univariate case, these intervals can and should be adjusted for bias. The bias adjusted form of the intervals uses the formulas

$$\bar{X}-\bar{Y}\pm\left(B+t_{df,1-\alpha/2}SE\left(\bar{X}-\bar{Y}\right)\right)$$

where B is bound on bias for the difference of the means. If bias bounds are known only for individual means—say, B_1 and B_2—then we may simply set $B=B_1+B_2$. Notice the bounds on bias are classically handled as additive, whereas random errors have standard errors that are the square root of the sum of the corresponding variances. There is one caveat: If both measurements are known to have equal bias, then the difference in means will have no bias.

We now present examples of these three tests and intervals for comparing two population means.

Example 7.2

The first example compares the travel time index (a measure of congestion) for eighty-six metropolitans for two years—1998 and 2003. Box plots for those data are shown in Figure 7.3.

A visual inspection of the box plot does not show much difference in the median of the two data sets. If a pooled or unequal variance t-test is performed on these data, there would be no significant difference found in the mean travel time index. The reason for the lack of power of these two t-tests is due to the high correlation of travel time index between the two years. This is shown clearly in Figure 7.4.

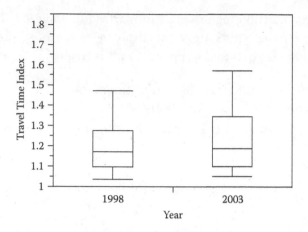

FIGURE 7.3   Box plots for the travel time index.

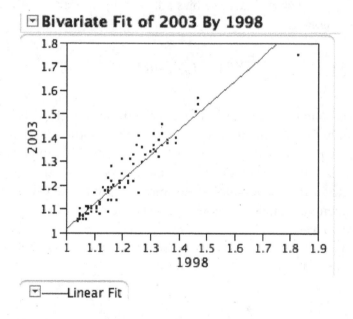

FIGURE 7.4   Scatter plot of the travel time index for the years 1998 and 2003.

Due to the fact that the travel time index is computed for the same cities for both years, the data are highly correlated. The proper t-test for this situation and data set is the paired t-test, and that output is shown in Figure 7.5.

Because the p value is <0.0001, the paired t-test rejects the hypothesis of no change in the mean travel time index ($p < .0001$). In other words, there is sufficient evidence in the data to conclude that there is a change in the mean

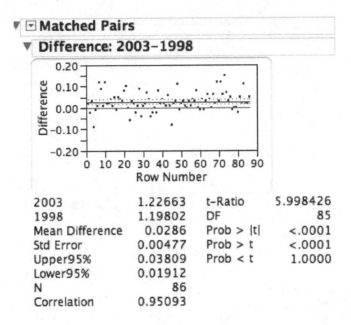

▼ ⊡ Matched Pairs

▼ Difference: 2003–1998

2003	1.22663	t–Ratio	5.998426
1998	1.19802	DF	85
Mean Difference	0.0286	Prob > |t|	<.0001
Std Error	0.00477	Prob > t	<.0001
Upper95%	0.03809	Prob < t	1.0000
Lower95%	0.01912		
N	86		
Correlation	0.95093		

FIGURE 7.5   Plot of difference and statistics for the paired *t*-test for mean travel times.

travel time index. The confidence interval (0.01912, 0.03809) that was output indicates that there is between a 2 and 4% increase in the mean travel time index during the five-year span.

In general, the paired *t*-test is often referred to as a stronger test than the pooled test. This is because much of the variability is internalized by using the same subject and only varying one factor. Consequently, considerably less data may be needed for the paired test, as compared to the pooled test, to have the same level of confidence. Unfortunately, because the requirements for conducting paired *t*-tests in transportation (e.g., only one parameter changes while all else remains the same) are comparatively rare, the reader will most likely work with pooled tests on a more consistent basis.

Example 7.3

To demonstrate the pooled and unequal variance *t*-tests, we test the hypothesis that the average speeds are equal on two different highway locations in Houston. A scatter plot and check of correlation (not shown) indicated that the speeds on the highway locations are reasonably modeled as independent. Box plots shown in Figure 7.6 indicate that the driving patterns on the two different road segments are similar.

The pooled *t*-test and unequal variance *t*-test both conclude that the null hypothesis of equal mean speeds cannot be rejected. This is shown in Figures 7.7 to 7.9. As long as the variances for the two populations are not

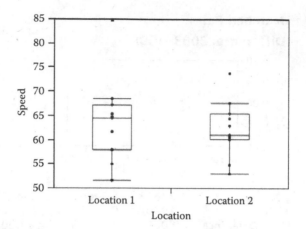

FIGURE 7.6   Box plots for speeds at two different locations.

t Test

location 2–location 1
Assuming equal variances

Difference	−1.3115	t Ratio	−0.60084
Std Err Dif	2.1827	DF	42
Upper CL Dif	3.0934	Prob > |t|	0.5512
Lower CL Dif	−5.7163	Prob > t	0.7244
Confidence	0.95	Prob < t	0.2756

−8 −6 −4 −2 0 2 4 6 8

FIGURE 7.7   Pooled *t*-test for mean speeds.

t Test

location 2–location 1
Assuming unequal variances

Difference	−1.3115	t Ratio	−0.60084
Std Err Dif	2.1827	DF	36.02118
Upper CL Dif	3.1152	Prob > |t|	0.5517
Lower CL Dif	−5.7381	Prob > t	0.7241
Confidence	0.95	Prob < t	0.2759

−8 −6 −4 −2 0 2 4 6 8

FIGURE 7.8   Unequal variance *t*-test for mean speeds.

▼ **Test Standard Deviation=value**

Hypothesized Value	4
Actual Estimate	3.98955
df	10

ChiSquare

Test Statistic	9.9478
Prob > |ChiSq|	0.8902
Prob < ChiSq	0.5549
Prob > ChiSq	0.4451

FIGURE 7.9   Test for population standard deviation of speed.

very different (e.g., say their ratio is within a factor of 3), then it is generally advisable to report the pooled t-test results. This is because the pooled t-test is more widely used and understood than the unequal variance t-test. However, when the variances are very different, then the separate variance t-test is the more appropriate of the two tests. Notice that the degrees of freedom for the two tests are, respectively, 42 and 36 (approximately).

7.5 INFERENCES ABOUT ONE POPULATION VARIANCE

Sometimes the transportation professional is not interested in tests about the central tendency of the data (e.g., sample mean) but rather of the dispersion, or variance. Similar to the above analysis, confidence intervals for the variance or standard deviation can be developed. Just as in the case of means, this book starts with simple hypotheses. The first set of hypotheses considered about variances is

$$H_0 : \sigma^2 = \sigma_0^2$$

$$H_a : \sigma^2 \neq \sigma_0^2.$$

The null hypothesis is that the population variance equals a specified value, while the alternative hypothesis is that the population variance does not equal the specified value. Recall from Chapter 6 that

$$\frac{(n-1)S^2}{\sigma^2} = \frac{\sum_{i=1}^{n}(X_i - \bar{X})^2}{\sigma^2} \sim \chi_{n-1}$$

if X_i s ($i = 1,...,n$) are a random sample selected from a *normal* population with mean μ and variance σ^2. Thus, under usual assumptions of normally distributed, identically distributed, and independent data, the appropriate tests and confidence intervals for a population variance are based upon a chi-square distribution.

The chi-square test rejects the H_0 when

$$\frac{(n-1)S^2}{\sigma_0^2} < \chi_{n-1,\alpha/2}^2$$

or when

$$\frac{(n-1)S^2}{\sigma_0^2} > \chi^2_{n-1,1-\alpha/2}.$$

That is, if the null hypothesis is true, we expect the ratio S^2/σ_0^2 to be close to 1, and if the ratio is either too small or too big, we reject the null hypothesis. The corresponding $(1-\alpha)100\%$ confidence interval for σ^2 is

$$\frac{(n-1)S^2}{\chi^2_{n-1,1-\alpha/2}} \leq \sigma^2 \leq \frac{(n-1)S^2}{\chi^2_{n-1,\alpha/2}}.$$

Corresponding tests and confidence intervals for the population standard deviation, σ, can be obtained by taking the square root of all sides of the above inequalities. These intervals may be bias corrected. Suppose that probe vehicles were used to measure speeds only during the summer when schools are not in session. That may cause a bias of $\pm B\%$. In that case, the bias corrected confidence interval is

$$\left(1 - \frac{B}{100}\right)\frac{(n-1)S^2}{\chi^2_{n-1,1-\alpha/2}} \leq \sigma^2 \leq \left(1 + \frac{B}{100}\right)\frac{(n-1)S^2}{\chi^2_{n-1,\alpha/2}}.$$

Example 7.4

In Figure 7.9 we presented the JMP output for testing the population standard deviation from a single location in Houston. The standard deviation is 4 mph based on a sample size of 11.

The corresponding confidence interval is displayed in Figure 7.10 along with the confidence interval for the mean speed. Notice that confidence intervals for standard deviations and variances are not centered at S^2. It can be seen that the null hypothesis (e.g., standard deviation is 4 mph) cannot be rejected in this case.

▼ Confidence Intervals

Parameter	Estimate	Lower CI	Upper CI	1-Alpha
Mean	60.71827	58.03806	63.39849	0.950
Std Dev	3.989547	2.787565	7.00139	

FIGURE 7.10 Ninety-five percent confidence intervals for mean speed and the standard deviation of speed.

7.6 INFERENCES ABOUT TWO POPULATION VARIANCES

The next situation that we consider is about testing the equality of two population variances. Let's consider the following hypotheses:

$$H_0 : \sigma_1^2 = \sigma_2^2$$

$$H_a : \sigma_1^2 \neq \sigma_2^2.$$

The usual test for these hypotheses is an F-test. The null hypothesis is rejected if the F-statistic is either too big or too small. We use the convention of writing the F-statistic with the larger of the two sample variances in the numerator, and the smaller of the two sample variances in the denominator. Then the traditional F-test rejects the null hypotheis only if the F-statistic is too large. Specifically, this is defined as

$$F = \frac{\max\left(S_1^2, S_2^2\right)}{\min\left(S_1^2, S_2^2\right)}.$$

Let the degrees of freedom for the sample variance estimate chosen for the numerator of the F-statistic be denoted as *numdf*, and the corresponding degrees of freedom for the variance estimate for the denominator be denoted as *denomdf*. The null hypothesis is rejected if $F > F_{numdf, denomdf, 1-\alpha}$. The JMP output for testing these hypotheses for equality of the variances of speeds at two locations is given in Figure 7.11.

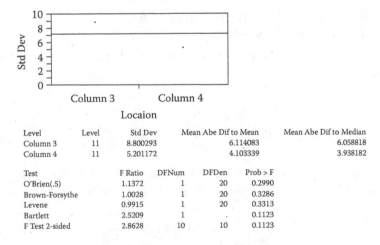

Level	Level	Std Dev	Mean Abe Dif to Mean	Mean Abe Dif to Median
Column 3	11	8.800293	6.114083	6.058818
Column 4	11	5.201172	4.103339	3.938182

Test	F Ratio	DFNum	DFDen	Prob > F
O'Brien(.5)	1.1372	1	20	0.2990
Brown-Forsythe	1.0028	1	20	0.3286
Levene	0.9915	1	20	0.3313
Bartlett	2.5209	1	.	0.1123
F Test 2-sided	2.8628	10	10	0.1123

FIGURE 7.11 Plot and tests for equality of variance from two populations.

The p value for this F-test is 0.11. Several other tests are also performed, and the first four shown in the list are useful when testing the equality of variances among $g \geq 2$ populations (see Chapter 8). Bartlett's test is very sensitive for departures from normality; hence, it may not be a good test if the data being tested have too many outliers. The other three tests—O'Brien, Brown-Forsythe, and Levene—are based upon ANOVA type procedures (that will be discussed in Chapter 8) with the dependent variable being a measure of spread. For example, Levene's test has the dependent variable as the absolute deviation of each observation from the group mean. A detailed explanation of these tests is beyond this text but may be found elsewhere (see, e.g., JMP, 2009; Boos and Cavell, 2004; Brown and Forsythe, 1974; and O'Brien, 1979).

7.6.1 Confidence Intervals for the Ratio of Two Variances

As with tests comparing two means, there are equivalent confidence intervals. These confidence intervals are for the ratio of variances. The null hypothesis of equal variances is rejected if they do not include the number 1. The confidence interval formula is

$$\frac{S_1^2}{S_2^2} F_L \leq \frac{\sigma_1^2}{\sigma_2^2} \leq \frac{S_1^2}{S_2^2} F_U.$$

Here $F_L = F_{\alpha/2, df1, df2}$ and $F_U = F_{1-\alpha/2, df1, df2}$ are the $\alpha/2$ and $1-\alpha/2$ quantiles (percentiles) from an F-distribution with $df1$ and $df2$ degrees of freedom.

7.6.2 Bias-Corrected Confidence Interval for the Ratio of Two Variances

If the bias in S_i^2, $i = 1,2$ might be $\pm B_i\%$, then the bias corrected confidence interval is

$$\left(\frac{1-B_1/100}{1+B_2/100}\right)\frac{S_1^2}{S_2^2} F_L \leq \frac{\sigma_1^2}{\sigma_2^2} \leq \left(\frac{1+B_1/100}{1-B_2/100}\right)\frac{S_1^2}{S_2^2} F_U.$$

Suppose that we knew that each variance estimate might be subject to a bias of ±5%. Then, the bias corrected form of the confidence interval is

$$\left(\frac{.95}{1.05}\right)\frac{S_1^2}{S_2^2} F_L \leq \frac{\sigma_1^2}{\sigma_2^2} \leq \left(\frac{1.05}{.95}\right)\frac{S_1^2}{S_2^2} F_U.$$

Note that the bias corrected form of the confidence interval is wider than the uncorrected form.

7.6.3 One-Sided Tests

Throughout the last two sections we have been testing two-sided hypotheses. Namely, the null hypothesis is that the variance was equal to a specific value (or two variances are equal), and the alternative hypothesis is that the variance is not equal to that value (or two variances are not equal). Sometimes we will want to test whether an experimental treatment improves performance measures. In those cases, we will use one-sided tests. For one-sided tests, the test statistics are based upon the ratio of the sample variances and the cutoff values change. See Ott and Longnecker (2009) for the appropriate one-sided tests.

7.7 CONCLUDING REMARKS

Transportation professionals are often asked questions about the way the system is functioning: Are cars speeding? Is a roadway built by a contractor within contract specifications? Needless to say, there is variability in almost all physical phenomena, and this randomness needs to be accounted for in order to answer these questions. This chapter introduced a number of important concepts to provide a statistical basis for these analyses.

The approaches in this chapter all make use of the sampling distributions discussed in Chapter 6, and two primary approaches were illustrated. The first were hypothesis tests that answered simple yes or no type questions related to the population mean or variance. The second were confidence intervals that looked at the likely bounds of the population mean or variance. Both techniques can be used to answer the questions or hypotheses related to a given metric, and if applied correctly, they will give the same result. However, it is the authors' contention that the confidence intervals provide more information to the analyst, and would be more useful in most engineering applications. In our experience, engineers often have a more intuitive grasp of the CI approach than the straightforward hypothesis testing.

Moreover, the analyst needs to know the basic assumptions before using these tests. As a simple example, if the samples are not random, then none of the approaches in this chapter apply. If there is bias in the measurements, this should be accounted for in the analysis. Common pitfalls were also illustrated; for example, almost any difference can be shown to be "statistically significant" if enough observations are taken. It is up to the analyst to determine whether a given difference is of practical importance—even if it is statistically significant.

The concepts developed in this chapter will be used throughout the remainder of this textbook and, in all likelihood, will be used throughout your career. It is imperative that the assumptions and theory are well understood before moving on to the next chapters.

APPENDIX: WELCH (1938) DEGREES OF FREEDOM FOR THE UNEQUAL VARIANCE *t*-TEST

Let

$$c = \frac{S_1^2}{S_1^2 + S_2^2},$$

the Welch degrees of freedom are

$$df^* = \frac{(n_1 - 1)(n_2 - 1)}{(1 - c)^2 (n_1 - 1) + c^2 (n_2 - 1)}.$$

As stated in Ott and Longnecker (2009), if the degrees of freedom are not an integer, then df^* should be rounded down to the nearest integer. The package JMP uses fractional degrees of freedom.

HOMEWORK PROBLEMS

1. A study is undertaken to examine the effects of drunk drivers and, in particular, whether older drivers are more affected by alcohol than younger drivers. Volunteers drink specified amounts of an alcoholic beverage at fixed time intervals. They perform baseline driving tests as well as driving tests fifteen minutes after each alcoholic drink. The age of each volunteer is recorded. Suppose that the outcome measure is the amount of time it takes for the drivers to complete a trip through an obstacle course. The recorded time is penalized if obstacles are missed or run over. What would be reasonable null and alternative hypotheses for this study?

2. Law enforcement wants to know typical driver speeds in front of a school when children are present. They also want to know if the average speed is exceeding the posted speed limit over the long term. The posted speed limit at this time is 35 mph. They use a radar gun that is accurate to within 1 mph (bias) and a standard deviation of 1 mph. The daily average speeds for a week are 37, 34, 38, 35, and 34 mph. With 95% confidence, is the speed limit exceeded (long term) on average based upon the data?

3. In the last problem, if the daily average speeds were misreported as 37, 34, 38, 35, and 84 mph, would the conclusion change? What is the effect of an outlier on the corresponding confidence interval and hypothesis test?

4. Probe vehicles are used to compute travel times of highway segments around San Antonio, Texas. Forty-five vehicles are used and they drive for approximately one hour at a time. A transportation center wants to compare the travel times on two nearby highway segments. Which of the following tests would be most appropriate and why: paired t-test, pooled t-test, or unequal variance t-test? What plot could you use to help confirm your answer?

5. Using the urban mobility data set, test whether the variances for public transportation passenger miles between 1982 and 2003 are the same for the two groups: large urban areas and very large urban areas.

6. Using the pavement marking data:

   a. Test the hypothesis that the variance for the detection distance for females is different from the variance for the detection distance for males.

   b. Recommend a confidence interval for the difference in detection distance for females and males. Would you use the equal or unequal variances assumption when constructing the confidence interval?

   c. Determine how the confidence interval would change if the device measuring the distance has a bias of 3 feet. (Hint: If a bathroom scale measures 1 pound overweight, how does that affect the difference in weight of two people who use the scale?)

REFERENCES

Ang, A. H.-S., and W. H. Tang. 2006. *Probability concepts in engineering: Emphasis on applications to civil and environmental engineering.* New York, NY: John Wiley & Sons.

Boos, D. D., and B. Cavell. 2004. Comparing variances and other measures of dispersion. *Statistical Science* 19:571–78.

Brown, M. B., and A. B. Forsythe. 1974. Robust tests for equality of variances. *Journal of the American Statistical Association* 69:364–67.

Knafl, G., C. Spiegelman, J. Sacks, and D. Ylvisaker. 1984. Nonparametric calibration. *Technometrics* 26:233–41.

Lehmann, E. L., and J. P. Romano. 2005. *Testing statistical hypothesis.* New York: Springer-Verlag LLC.

O'Brien, R. G. 1979. A general ANOVA method for robust test of additive models for variance. *Journal of the American Statistical Association* 74:877–80.

Ott, R., and M. Longnecker. 2009. *An introduction to statistical methods and data analysis.* 6th ed. Belmont, CA: Cengage/Brooks/Cole.

SAS Institute. JMP 2009. Version 8. Cary, NC: SAS Institute.

CHAPTER 8

Other Inferential Procedures

ANOVA and Distribution-Free Tests

8.1 INTRODUCTION

Transportation researchers often have to compare many possible intervention strategies. For example, they may question whether it is better to use a sign that reads "slower traffic keep right," a sign that reads "left lane for passing," or some combination of both. A natural solution may be to compare different combinations of these options and examine which gives the better result. It is the transportation analyst's role to compare the results of the different combinations. In this chapter, we explain analysis of variance (ANOVA), an often-used technique for comparing means from several populations. This technique allows the analyst to test various hypotheses, such as whether the mean response for the different combinations is statistically the same. If the hypothesis of equal means is rejected by the ANOVA test, it is natural to question which treatment, or signing strategy, is better. In this chapter we provide multiple comparison procedures that can be used to clarify these issues.

Many of the tests described in this textbook depend upon normality. While many phenomena in engineering behave in normal (or near-normal) ways, there are many situations in which this is not the case. This chapter also presents methods for testing the validity of the normality assumption and, when the assumption is not valid, presents an alternative method for testing equality of means.

8.2 COMPARISONS OF MORE THAN TWO POPULATION MEANS

In many situations transportation professionals need to compare the means of more than two populations, and this section presents two ways of accomplishing this task. The first way is by using confidence intervals (CIs), and the second is by using analysis of variance (ANOVA). When transportation professionals need to compare g-means, such as when $g \geq 2$, then there are $g(g-1)/2$ ways to compare pairs of means. For example, if $g = 3$, we would need to compare $3 \times 2/2 = 3$ pairs: (μ_1,μ_2), (μ_1,μ_3), and (μ_2,μ_3). In order to ensure that the type I error is no more than α, each confidence interval is constructed at the

$$\left(1-\frac{2\alpha}{g(g-1)}\right)\times 100\% \text{ level.}$$

This correction to α is known as a Bonferroni approach. It is widely used if g is not too big, typically for g less than or equal to 10. If g is large, then the confidence intervals frequently are too wide to be useful. Note that the confidence intervals can be calculated for either the equal or unequal variance scenarios. In addition, the test statistics used in the various tests need not be independent.

The ANOVA approach is a generalization of the pooled t-test that we used when $g = 2$ and we assumed that variances for the two populations were equal. The idea underlying ANOVA is a comparison of a pooled variance estimate to an estimate of variance obtained using the individual sample means.

Suppose the hypotheses to be tested are

$H_0 : \mu_1 = \mu_2 = \ldots = \mu_g$, in which all means are equal

vs.

H_a : not all means are equal.

We also assume that the data are independently normally distributed with the same variance across the g populations. For pedagogical purposes only, let us assume that the number of observations from each population is equal. That is, we assume that there are $n_1, n_2, \ldots, n_g$ observations from each of the g populations and that $n_0 = n_1 = n_2 = \ldots = n_g$. The ANOVA F-test is the ratio of two quantities. The denominator for the F-test is a pooled variance estimate

$$\frac{(n_1 - 1)S_1^2 + (n_2 - 1)S_2^2 + \ldots + (n_g - 1)S_g^2}{\sum_{i=1}^{g} (n_i - 1)} = \frac{\left(S_1^2 + S_2^2 + \ldots + S_g^2\right)}{g}.$$

The numerator is the sample variance of the means multiplied by n_0. That is, the numerator equals

$$\left(\frac{n_0 \sum_{i=1}^{g} (\bar{X}_i - \bar{X})^2}{g - 1} \right).$$

Under the null hypothesis, both the numerator and the denominator have their expected value, σ^2. Thus, under the null hypothesis an F-test statistic is anticipated to be 1, and approximately 95% of the time would not provide enough evidence to reject the null hypothesis. The usual formulas used for ANOVA are given in almost every standard textbook; see, for example, Ang and Tang (2006) and Ott and Longnecker (2009). In our pedagogical context the numerator sum of squares, called the between sum of squares or the model sum of squares, equals

$$n_0 \sum_{i=1}^{g} (\bar{X}_i - \bar{X})^2$$

and its degrees of freedom are $g - 1$. The corresponding denominator sum of squares is called the within sum of squares or error sum of squares and is equal to

$$(n_1 - 1)S_1^2 + (n_2 - 1)S_2^2 + \ldots + (n_g - 1)S_g^2.$$

Additionally, it has $\sum_{i=1}^{g} (n_i - 1) = g(n_0 - 1) = n - g$ degrees of freedom. The mean within sum of squares is

$$S_w^2 = \frac{(n_1 - 1)S_1^2 + (n_2 - 1)S_2^2 + \ldots + (n_g - 1)S_g^2}{n - g}.$$

To demonstrate how confidence intervals and ANOVA can be used to compare means from several populations, an example from San Antonio, Texas, is used. Speed data from four traffic corridors, each with zero, one, two, or three traffic signals, was collected. Note that these data are a subset of the data used for Figure 2.3. The numbers of observations from each population are not equal, but JMP and other statistical programs easily handle this situation. It can be easily calculated that there are $4 \times 3/2 = 6$ comparisons to be made. Figure 8.1 shows the results of the six pairwise confidence intervals. The intervals are at the $(1-\alpha/6)100\% \approx 99.2\%$ level, and use the mean squared error as the estimate of the common variance. Thus, these tests differ from the typical pooled t-test because the variance estimate uses data from all the populations. In addition, the α level is properly adjusted to account for multiple comparisons.

It is evident in Figure 8.1 that the first five confidence intervals do not contain zero, but the sixth does. In this case, the population mean speeds for all signal conditions are different with the exception of corridors with one or two traffic signals. The ANOVA information shown in Figure 8.2 indicates that the hypothesis of equal population means can be rejected. That is, at least one of the means is different from the others. However, unlike the CI test, the ANOVA table does not identify the means that are

Number of Signals	Number of Signals	Difference	Standard Error Difference	Lower CL	Upper CL
0	3	17.89178	1.172762	15.5873	20.19625
0	1	15.12759	1.110275	12.9459	17.30927
0	2	15.00113	0.990213	13.0554	16.94689
2	3	2.89065	1.057042	0.8136	4.96773
1	3	2.76419	1.170268	0.4646	5.06376
2	1	0.12646	0.987258	−1.8135	2.06642

FIGURE 8.1 Bonferroni adjusted confidence intervals for four signaling levels.

Source	DF	Sum of Squares	Mean Square	F-Ratio	Probability > F
No. signals	3	20,734.016	6,911.34	106.2889	<.0001*
Error	473	30,756.379	65.02		
C. total	476	51,490.395			

FIGURE 8.2 ANOVA for speed data at four signaling levels.

likely to be unequal. For this reason, the authors argue (similar to that of Chapter 7) that CIs are the preferred method of analysis for this type of comparison.

8.3 MULTIPLE COMPARISONS

The experiment-wise type I error rate is the chance that any null hypothesis under consideration is falsely rejected. When the factor is under test, for instance, the drivers of probe vehicles have many levels, a multiple testing issue arises. While the type I error for ANOVA correctly sets the type I error at α, it does not tell us which levels are responsible for the rejection of the null hypothesis. Once we try to assert which pairs of factor levels are different, the α level is no longer controlled by ANOVA. We have already encountered one way around this issue. The Bonferroni method correctly limits the experiment-wise type I error to α. It is, however, less powerful than ANOVA at rejecting the null hypothesis when it is false. This means that even when true deviations from the null hypothesis are fact, the Bonferroni approach may not detect the deviations.

A popular approach is to use Fisher's least significant difference (LSD) procedure. This test uses all possible pairwise tests for equality of means. These tests would be the same as the pairwise pooled variance t-tests except that they use the mean sum of squares within to obtain S_w rather than using S_p. Thus, these tests have the larger $n-g$ degrees of freedom rather than the smaller $n_i + n_j - 2$ for the ijth comparison. It is wise not to do these tests unless the ANOVA test rejects the null hypothesis of equal means. Refer to Ott and Longnecker (2009) for additional details.

Another popular procedure is called Tukey's honestly significant difference (HSD) procedure. Readers interested in this topic should see Tukey (1953) for more information. These comparisons are based upon the normalized differences:

$$\frac{\left|\bar{X}_i - \bar{X}_j\right|}{\sqrt{S_w^2 / n}}.$$

The HSD critical value q depends upon g (the number of levels of the factor) and the degrees of freedom of S_w. This test is available in JMP, and it should be noted that ANOVA does not have to be a precursor for the HSD method to be applied. This test correctly accounts for multiple comparisons on an experiment-wise basis. Again more details about this method can be found in Ott and Longnecker (2009).

To illustrate these methods, we reanalyze the data from San Antonio used in Figure 2.3. The dependent variable is average speed along the corridor, and the factor that we use is the number of traffic signals on the corridor. We present the JMP output in Figure 8.3.

The first observation that can be made from the output of Figure 8.3 is that even small differences in the populations can be detected as statistically significant due to the relatively large sample sizes. From Fisher's LSD procedure (given under "Comparisons for each pair using Student's *t*" in Figure 8.3) we can see that only the six and eight signal routes have indistinguishable speeds. This can be deduced since the populations for six and eight share a common letter. On the other hand, from the Tukey HSD (given under "Comparisons for all pairs using Tukey–Kramer HSD" in Figure 8.3), it is evident that populations 3 and 4 share a common letter, which is also true of populations 6 and 8. The three- and four-signal routes have indistinguishable speeds, and the six- and eight-signal intersections are not statistically different.

8.4 ONE- AND MULTIWAY ANOVA

The ANOVA method used to analyze the speed vs. signal data is called one-way ANOVA. That is because only one factor or independent variable, the number of signals, was used in the analysis. Speed is the dependent variable. Intuitively, there are likely many more independent variables that affect speed than simply the number of traffic signals. In this situation a multiway ANOVA would be required for analysis. The underlying model for one-way ANOVA is

$$Y_{ij} = \mu + \mu_j + \varepsilon_{ij}, i = 1,\ldots,n_j, j = 1,\ldots,g.$$

The observed measurements are denoted by Y_{ij}. For pedagogical purposes, we will assume that $n_1 = n_2 = \ldots = n_g = n_0$, and so $n = g \times n_0$ is the total number of measurements. In order for the parameters to logically correspond, it is common to assume that the parameter μ is the overall mean with the expected value taken over all groups. The parameters $\mu_j, j = 1,\ldots,g$, are the mean offset for the *j*th group from the overall mean. It is often assumed that $\sum_{j=1}^{g} \mu_j = 0$. The random errors $\varepsilon_{ij}, i = 1,\ldots,n_j, j = 1,\ldots,g$, are assumed independently normally distributed with a mean of zero and constant variance of σ^2. The degrees of freedom for the one-way ANOVA model sum of squares are g, and the error sum of squares has $n - g$ degrees of freedom.

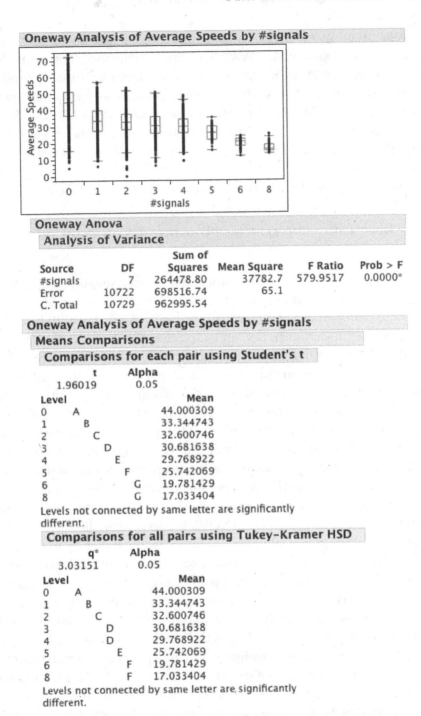

FIGURE 8.3   ANOVA and multiple comparisons for speed vs. number of signal data from San Antonio.

When there is more than one factor, ANOVA extends to multiway ANOVA. In two-way ANOVA there are two factors. For example, the number of traffic signals on a road segment and weather conditions are factors that may affect driving speed. A new effect emerges for multiway ANOVA, and is referred to as interaction effects. These effects account for more changes in mean level where factors act differently together than they would if their effects were independent of each other.

The basic two-way model is written as

$$Y_{ijk} = \mu + \mu_{i.} + \mu_{.j} + \gamma_{ij} + \varepsilon_{ijk}, \; i = 1, \ldots, I, \; j = 1, \ldots, J, k = 1, \ldots, n_{ij}.$$

Similar to the one-way case, the observable measurements are Y_{ijk}. Once again, for pedagogical purposes, we will assume that $n_{ij} = n_0$, $i = 1, \ldots, I$, $j = 1, \ldots, J$. The parameter μ represents the mean level across all levels of both factors. The parameters $\mu_{i.}$ and $\mu_{.j}$ represent the mean offsets from μ for the effect of the first and second factors, respectively. These are called main effects, and they are interpreted as the part of the factor effects that are unaffected by interactions. It is often assumed that $\sum_{i=1}^{I} \mu_{i.} = 0$ and $\sum_{j=1}^{J} \mu_{.j} = 0$. The parameters γ_{ij} represent the interaction effects among the ith level of the first factor and the jth level of the second factor. The parameters γ_{ij} account for the lack of additivity of the effects for the first and second factors. For example, rainy weather may slow traffic more when there are three traffic signals along the corridor in comparison to when there are no traffic signals on a corridor. These interaction terms will be analyzed as part of the ANOVA test to see whether the hypotheses can be supported by the data. It is often assumed that $\sum_{i=1}^{I} \gamma_{ij} = 0$ and $\sum_{j=1}^{J} \gamma_{ij} = 0$. The constraints are needed because without such restrictions there are more parameters than possible levels of the mean response. The constraints are a natural way to make the parameters identifiable. Readers should see Ott and Longnecker (2009) for a more detailed explanation of this concept. Finally, the random variables ε_{ijk}, $i = 1, \ldots, I$, $j = 1, \ldots, J$, $k = 1, \ldots, n_{ij}$, are assumed to be independent normal random variables with zero mean and common variance. The ANOVA table gives $I - 1$, $J - 1$, and $(I - 1)(J - 1)$ degrees of freedom for the first and second factors, as well as interaction effects. The error degrees of freedom are $n - (I - 1) - (J - 1) - (I - 1)(J - 1)$. For a more in-depth discussion see Ott and Longnecker (2009) and Kuehl (2000).

As an example, let us consider the San Antonio average speed data with weather and signals as factors. By way of demonstration we will use four levels of weather (denoted 1, 2, 3, and 4) and four levels of corridor (zero, one, two, and three traffic signals). The results of a two-way ANOVA for these data are shown in Figure 8.4.

From the ANOVA table in Figure 8.4 we can see that both main and interaction effects are significant. Not surprisingly, weather affects travel speed, as do the number of traffic signals. Furthermore, the interaction of these two factors further affects travel speed. All the p values in the ANOVA are less than .0001. That is much less than the customary type I

(a) Response Average Speeds
Summary of Fit

RSquare	0.252302
RSquare Adj	0.251138
Root Mean Square Error	8.197914
Mean of Response	34.59409
Observations (or Sum Wgts)	9656

Analysis of Variance

Source	DF	Sum of Squares	Mean Square	F Ratio
Model	15	218614.01	14574.3	216.8603
Error	9640	647863.81	67.2	Prob > F
C. Total	9655	866477.82		0.0000*

Effect Tests

Source	Nparm	DF	Sum of Squares	F Ratio	Prob > F
Weather	3	3	5019.6797	24.8971	<.0001*
#signals	3	3	1618.7689	8.0289	<.0001*
Weather*#signals	9	9	5516.4017	9.1202	<.0001*

Effect Details

Weather*#signals
Least Squares Means Table

Level	Least Sq Mean	Std Error
1,0	44.249248	0.2483076
1,1	32.707129	0.2250442
1,2	32.718416	0.1637619
1,3	30.801912	0.2528727
2,0	43.960328	0.2969791
2,1	34.481262	0.3070143
2,2	32.695848	0.2596305
2,3	30.630899	0.3003483
3,0	31.968529	1.4059306
3,1	33.513370	0.8546916
3,2	30.741885	0.5931802
3,3	29.820763	0.7546791
4,0	37.280000	8.1979137
4,1	39.020000	4.0989569
4,2	31.584000	2.1166922
4,3	31.285000	2.5924079

FIGURE 8.4 Two-way ANOVA with effect details for San Antonio speed data.

(b) LS Means Plot

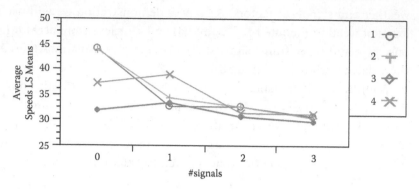

LSMeans Differences Tukey HSD
α=0.050

Level					Least Sq Mean
1,0	A				44.249248
2,0	A				43.960328
4,1	A	B	C	D	39.020000
4,0	A	B	C	D	37.280000
2,1		B			34.481262
3,1		B	C	D	33.513370
1,2			C		32.718416
1,1			C		32.707129
2,2			C		32.695848
3,0		B	C	D	31.968529
4,2		B	C	D	31.584000
4,3		B	C	D	31.285000
1,3				D	30.801912
3,2			C	D	30.741885
2,3				D	30.630899
3,3				D	29.820763

Levels not connected by same letter are significantly different.

FIGURE 8.4 (Continued) Two-way ANOVA with effect details for San Antonio speed data.

error rate of .05. It should be noted that when the interaction effects are significant, it may not make much sense to discuss the main effects (without referring to the level of the other factor) because the effect of one factor may be different for each level of the other factor. Our recommendation is that whenever an ANOVA test suggests significant interactions, the interaction plots need to be examined first. An interaction plot is a graph of predicted mean responses that are connected line segments. In Figure 8.4, the interaction plot for weather and number of signals is given under "LS Means Plot." The Y-axis is the predicted mean response, and the X-axis displays the levels of the number of signals factor. Separate line segments are drawn for each level of the other factor, weather, as a function of all levels of the factor on the X-axis.

If there is no interaction between the plotted factors, then all the line segments will be parallel since main effects are additive. If there is an important interaction effect, the line segments will noticeably not be parallel, as the effects of factors differ with the combination of levels. If the connected line segments are far from being parallel, like the one in Figure 8.4, the assessment of the effects of the factors needs to be made based on the interaction plots and the corresponding least-squares means, rather than on the results from the tests of main effects. For example, the interaction plot of Figure 8.4 indicates that the effect of weather on travel speed is different depending on the level of the number of traffic signals. It is evident that weather affects travel speed when there is zero or one traffic signal, whereas weather is irrelevant for two or three traffic signals. Put another way, when the level of weather is 1 or 2, the average travel speed is higher when there is no signal compared to when there is a signal, but when the level of weather is 3 or 4, the average travel speed is lower when there is no signal compared to when there is one signal. (Tukey's procedure does not indicate a significant difference in the average travel speed when the number of signals is two or three regardless of the level of weather.) The multiple comparison tests can be used to determine which of the least-squares means of the interaction plots are statistically different, and are given under "LS Means Differences Tukey HSD."

Remark 8.1

In some cases, the interaction effects may not be practically significant, although they are statistically significant (or the line segments of the interaction plots may not be far from being parallel). In such cases, the interactions could be assumed to be negligible and the assessment of main effects is still logical.

8.5 ASSUMPTIONS FOR ANOVA

The tests in the previous sections are all predicated on the underlying assumptions for ANOVA, which are normality, independence, and an equal variance. ANOVA is robust to the extent outliers make it difficult to reject the null hypothesis when it is true. Unfortunately, outliers also make it challenging to reject the null hypothesis when it is false. The constant variance assumption can be checked using the Levene test that was introduced in Chapter 7. When the equal variance assumption is provably wrong (i.e., when the p value is less than α and the difference in variances is practically significant), the unequal variance test that is available in JMP should be used. ANOVA is reasonably robust to moderate differences in variances. Completing a standard ANOVA and comparing it to an unequal variance ANOVA can indicate the effects of unequal variances. Finally, the assumption of independence is very important. If there is reason to suspect a dependence in the measurements over time—for example, the same cars have their speeds measured on different links—then professional statistical assistance should be sought. Techniques such as repeated measure designs are beyond the scope of this textbook. However, a time series plot, as described in Chapter 2, can be helpful for detecting dependencies among the data.

The most common way to check the assumption of normality is to plot the residuals, which are the differences between the estimated model and the observations (see Chapter 10 for more discussion on the residual plot). For example, in the two-way ANOVA case, the residuals would be

$$r_{ijk} = Y_{ijk} - \hat{\mu} + \hat{\mu}_{i.} + \hat{\mu}_{.j} + \hat{\gamma}_{ij}, i = 1,\ldots,I, j = 1,\ldots,J, k = 1,\ldots,n_{ij}.$$

If upon analysis of the residuals they do not appear to be approximately normally distributed, it is common to try to transform the data. Before the use of computer programs, this could be a time-consuming process, as the analyst would have to identify a promising transformation and apply the formula. Then, the analyst would need to estimate the model and plot the residuals to see if the transformation resulted in normal data. Most modern statistical packages provide an automated transformation function that will find an optimal transformation to use for fitting the data. In the case of JMP it is based on the Box-Cox transformation, which is further explained in Box and Cox (1964). By applying it, JMP will find the optimal transformation to use in fitting the data. The form of the Box-Cox transformation is

$$\frac{Y^{\lambda} - 1}{\lambda(GM)^{\lambda-1}}$$

for $\lambda \neq 0$ and $\ln(Y)$ for $\lambda \approx 0$. The term $GM = (Y_1 Y_2 \ldots Y_n)^{1/n}$ is the geometric mean. It is common that if $\lambda \approx 0$, use the log transform, and if $\lambda \approx .5$, use the square root transform, while if $\lambda \approx 1$, use no transform.

Figure 8.5 contains the normal quantile plot (see Chapter 2) of the residuals that can be used for assessing the assumption of normality. From

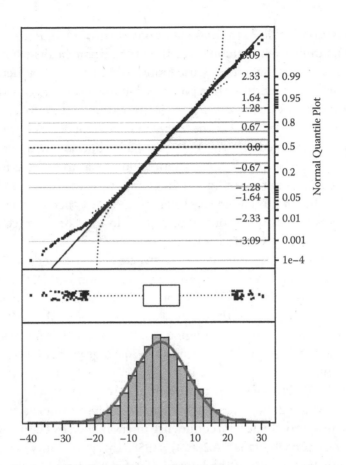

FIGURE 8.5   Residual plots for the fit of weather and signals to speed data.

Figure 8.5 we can see that the normal assumption for errors is not severely violated, but there are a good number of outliers. The effects of the outliers/ violation of normality assumptions may be different from case to case. To assess such effects, one can either transform the data (average speeds) before fitting ANOVA, or use alternative tests that do not depend on the normality assumption (such as those presented in the next section). Then, one should compare the results against those from the original ANOVA. A homework problem is given at the end of this chapter for such a comparison.

8.6 DISTRIBUTION-FREE TESTS

So far, when we tested hypotheses about means, the tests assumed that distributed data were essentially normal. From a practical perspective, we assumed that there were no huge outliers and there were no more outliers than would be expected from normally distributed data. In many transportation problems these assumptions are invalid. In this section, we present a number of commonly used nonparametric or distribution-free testing procedures that are less susceptible to the effects of outliers.

We start by testing the null hypothesis that the median of a distribution is some specified value or not. As we stated earlier, the sample median is not as sensitive to outliers as the sample mean. Suppose that we want to test whether the median of a probability distribution equals m_0 vs. the alternative hypothesis that the median is not equal to m_0. We form this test by constructing the confidence interval for the median and checking whether m_0 lies in this confidence interval. If it does, then we do not reject the null hypothesis. On the other hand, if m_0 does not lie in the confidence interval, we reject the null hypothesis.

The confidence interval will be constructed using order statistics. We assume that we have a random sample of observations $X_1, X_2, \ldots, X_N$ from a continuous distribution. Let the ordered (sorted) data be denoted as $X_{(1)} < X_{(2)} < \ldots < X_{(n)}$. Thus, the smallest observation from the random sample is denoted as $X_{(1)}$ and the largest as $X_{(n)}$. The confidence interval for the median has the form $[X_{(k)}, X_{(n-k+1)}]$, and the coverage probability associated with this interval is $1-\alpha$, where $\alpha = 2 \sum_{i=0}^{k-1} \binom{n}{i} (1/2)^n$. We assume that k is less than n/2. Notice that α takes only a few discrete values as k ranges between 1 and $n/2$. Suppose that we have a random sample size of 30. If we want an approximate 95% confidence interval, then by choosing $k = 10$, the coverage probability is $1 - 2(.021) = .958$. This probability can be computed with JMP by using the binomial distribution within the formula tool for columns. As in the case of testing one population mean, there are distribution-free tests for comparing two or more means. Readers are referred to Hollander and Wolfe (1999).

8.6.1 The Kolmogorov–Smirnov Goodness-of-Fit Test

The empirical distribution function is defined as

$$\hat{F}_n(x) = \frac{\text{\# of observations} \leq x}{n}.$$

In a large sample it effectively approximates the true distribution function. We initiate the Kolmogorov–Smirnov test by stating that for any fixed sample size, n, and empirical distribution function, $\hat{F}_n(x)$, the term $n\hat{F}_n(x)$ has a binomial distribution with parameter n, $p = F(x)$. This implies that the mean of $\hat{F}_n(x)$ is $F(x)$ and its variance is $F(x)(1-F(x))$. We use these facts to construct a confidence interval for $F(x)$. An approximate $100(1-\alpha)\%$ confidence interval is

$$\hat{F}_n(x) \pm z_{1-\alpha/2}\sqrt{\hat{F}_n(x)(1-\hat{F}_n(x))/n}.$$

This confidence interval can be used to test whether the true distribution of the data comes from distribution $F_0(x)$. That is, we can test the hypotheses

$$H_0 : F(x) = F_0(x)$$

$$H_a : F(x) \neq F_0(x)$$

by checking whether or not $F_0(x)$ lies in the confidence interval $(1, 1)$. If it does, we cannot reject the null hypothesis, but otherwise, we can reject the null hypothesis.

Readers will notice that the hypotheses that we tested are particular to a point: x. If we want to test the hypotheses that

$$H_0 : F(x) = F_0(x)$$

$$H_a : F(x) \neq F_0(x)$$

for x we need another method. The Kolmogorov–Smirnov test is a popular method for such a test. It gives a constant, c, such that if the null hypothesis is true, then the set

$$\left[\hat{F}_n(x) - c\sqrt{\hat{F}_n(x)(1-\hat{F}_n(x))/n}, \hat{F}_n(x) + c\sqrt{\hat{F}_n(x)(1-\hat{F}_n(x))/n}\right]$$

(over all x) is a $100(1-\alpha)\%$ confidence interval for $F_0(x)$. The Kolmogorov–Smirnov test can be found within JMP's distribution platform (see the JMP manual for details). JMP gives a p value for this test rather than the corresponding confidence interval. Explanations of the other tests used by JMP can be found in Hollander and Wolf (1999).

8.6.2 The Kruskal–Wallis Approach to ANOVA

The lack of normality and, in particular, the presence of outliers in ANOVA lead to low power. Namely, the hypothesis of equal means is difficult to reject when there are many outliers because they inflate the error sum of squares. An alternative is to use a rank test in which all the data are placed in one column and ranked. The smallest data point gets rank 1, the next smallest rank 2, ..., and the largest is ranked as $n = n_1 + n_2 + ... + n_g$. Then the raw data are replaced by the ranked data and ANOVA is performed on the latter. This is known as the Kruskal–Wallis test, and it is very resilient to the presence of outliers. The description here is not the exact Kruskal–Wallis test, but a slight variant. Please refer to Kruskal and Wallis (1952) and Ott and Longnecker (2009) for further details. JMP calls this a Wilcoxon test and specifies that it is equivalent to the Kruskal–Wallis test when there are two or more groups.

8.7 CONCLUSIONS

This chapter illustrated some of the popularly used tests for comparing multiple means or situations when the common normality assumptions do not hold. These tests often occur in transportation applications, and a typical transportation professional will conduct, or at least be exposed to, such procedures on a regular basis. In these situations the ANOVA or the nonparametric version, the Kruskal–Wallis procedure, is an appropriate technique for comparing across the different levels.

Lastly, many of the tests rely on underlying assumptions of normality and assume that there are not a large number of outliers. Unfortunately, these situations are not uncommon in transportation engineering. Fortunately, there are a number of nonparametric tests that can be used to address this issue. As the types of questions that transportation professionals need to answer become more complex, these techniques will become more common to the profession.

HOMEWORK PROBLEMS

1. Suppose that we wanted to compare the average speeds among five segments for the probe vehicles. How can we modify 95% confidence intervals so that their combined error rate is 5% while accounting for all $5 \times 4/2 = 10$ comparisons?

2. Using the urban mobility analysis data, test the hypothesis that the mean values for public transportation passenger miles between 1982 and 2003 are the same for Akron, Ohio; Albuquerque, New Mexico; Allentown, Pennsylvania; and Anchorage, Alaska. If the means are not equal, form a grouping for these cities' public transportation means that is logically sound.

3. For the data used in problem 2 show that the mean squared error equals $S^2_{Akron} + S^2_{Albuquerque} + S^2_{Allentown} + S^2_{Anchorage} / 4$. If the sample sizes were not equal, would this be true? How would the formula change?

4. For pavement marking data the gender and glare variables were combined.

$$\begin{cases} 1 = \textit{Female, Glare} \\ 2 = \textit{Male, } \sim \textit{Glare} \\ 3 = \textit{Female, } \sim \textit{Glare} \\ 4 = \textit{Male, Glare.} \end{cases}$$

a. Do an ANOVA of the detection data using the combined gender and glare variable as the factor.

b. Do an ANOVA of the logged detection data using the combined gender and glare variable as the factor.

c. Which analysis is more appropriate using the logged data or the raw data? Why?

5. For the speed, number of signals, and weather data used for ANOVA, redo the analysis using rank transforms.

a. Are the main effects and interactions significant?

b. How do the p values compare to the p values from ANOVA?

REFERENCES

Ang, A. H.-S., and W. H. Tang. 2006. *Probability concepts in engineering planning and design: Basic principles.* Vol. 1. New York: John Wiley & Sons.

Box, G. E. P., and D. R. Cox. 1964. An analysis of transformations. *Journal of the Royal Statistical Society B* 26:211–52.

Hollander, M., and D. A. Wolf. 1999. *Nonparametric statistical methods.* Wiley Series in Probability and Statistics, 2nd ed. New York: Wiley.

Kruskal, W. H., and W. A. Wallis. 1952. Use of ranks in one-criterion variance analysis. *Journal of the American Statistical Association* 47:583–621.

Kuehl, R. O. 2000. *Design of experiments: Statistical principles of research design and analysis.* 2nd ed. Pacific Grove, CA: Duxbury.

Ott, R., and M. Longnecker. 2009. *An introduction to statistical methods and data analysis*. 6th ed. Belmont, CA: Dengage/Grooks/Cole.

Tukey, J. W. 1953. The problem of multiple comparison. In *The collected works of John W. Tukey VIII. Multiple comparisons: 1948–1983*. New York: Chapman & Hall.

CHAPTER 9

Inferences Concerning Categorical Data

9.1 INTRODUCTION

In Chapter 3, we introduced the concept of proportion(s) as a summary of categorical data. As you may recall, this type of data occurs relatively often in transportation. For example, the U.S. Federal Highway Administration (FHWA) groups vehicles into thirteen distinct types, and a good portion of survey data is categorical in nature. Consequently, transportation engineers use and test proportions quite often. We may wish to know if an education program decreases the percentage of teen drivers who are texting while driving. In another situation, we may want to know if there are differences in the proportions of male and female drivers using toll roads. Not surprisingly, it is often not enough to know what the proportions are for the questions in which we are interested; thereby, we would like to know how confident we are in the estimates. The focus in this chapter is on making inferences concerning population (true) proportions for categorical data; for example, it may be that 47% of male drivers and 53% of female drivers use toll roads. In this chapter we will learn how to infer whether the 6% difference is statistically significant.

9.2 TESTS AND CONFIDENCE INTERVALS FOR A SINGLE PROPORTION

Suppose we have a situation where each of n individuals has an independent choice. For example, drivers on State Route (SR) 91 in California have a choice of driving on either a toll or free section of the roadway—both sections run parallel to each other. Suppose we use notation

$$X_i = \begin{cases} 0 \text{ if option 1 is made} \\ 1 \text{ if option 2 is made} \end{cases}, \quad i = 1,\dots,n.$$

For our example, 0 may correspond to a driver using the free lanes and a 1 would correspond to a driver using the tolled lanes. Let $\hat{P} = \sum_{i=1}^{n} X_i / n$ and $p = E\hat{P}$. In an analogous approach to testing whether a sample mean equals μ or not, we will test whether $p = p_0$ or not. With a couple of technical caveats that we will discuss later, the philosophy of testing proportions is analogous to those for testing means from a continuous random variable. One caveat is that we must have a minimum number of observations in order to make meaningful confidence intervals. In this textbook we will assume that there are at least five successes (ones) and at least five failures (zeros) in any given trial.

As an aside, the rule about the number of successes is based upon traditional advice where np and $n(1-p)$ should be greater than 5. In essence, we are hoping that the central limit theorem (CLT) applies, and based on experience, this appears to be a good rule of thumb. However, in statistics it is critical to always check your assumptions.

Because we do not know p and it appears in the standard deviation formula that we use to create a confidence interval, we use $\hat{p}$ as a surrogate. We reject the null hypothesis $p = p_0$ when $\hat{p}$ is far from p_0 relative to its standard error. The estimated standard error of $\hat{p}$ is typically given as

$$\sqrt{\frac{\hat{P}(1-\hat{P})}{n}},$$

and in most transportation applications, this approximation will suffice. Therefore, the standard testing setup becomes

$$H_o : p = p_0$$

$$H_a : p \neq p_0$$

and the rejection rule is to reject the null hypothesis when

$$\frac{\sqrt{n}\left|\hat{P} - p_0\right|}{\sqrt{p_0(1-p_0)}} \geq z_{1-\alpha/2}. \tag{9.1}$$

Under the null hypothesis the standard error of $\hat{P}$ is

$$\frac{\sqrt{p_0(1-p_0)}}{\sqrt{n}}. \tag{9.2}$$

Note that for proportions, the typical confidence interval has a different form than you have previously seen due to the fact that there is no particular value for p (such as p_0) that is of central concern. Thereby, the confidence interval has the form

$$\hat{P} \pm \frac{z_{1-\alpha/2}\sqrt{\hat{P}(1-\hat{P})}}{\sqrt{n}}. \tag{9.3}$$

Also note that bias correction can also be added in the case of the confidence intervals for the mean. For example, if 2% of respondents to a survey leave a question blank, then ±2% can be added to both sides of the confidence interval.

As we discussed previously, there are two technical caveats that may have to be addressed. The first is when a given experiment results in observations that are either all successes or all failures. Note that the assumption of $np \geq 5$ and $n(1-p) \geq 5$ can hold and we can still not observe a success in ten trials. The reader will quickly see that the estimated standard error would be zero. For example, suppose that we are concerned with red light violations and we observed no red light offenders in ten cycles at an intersection. Then the usual 95% confidence interval for the proportion of cycles where red light offenses occur would be 0 ± 0. However, from a practical point of view this is a ridiculous answer. For this reason, the authors recommend a minimum of five successes and five failures before using these techniques.

An alternative approach is to estimate the intervals based upon calculations that use the binomial distribution because the binomial confidence interval can cope with the low number of observations. When $\hat{P} = 0$, the confidence interval is $(0, -(\alpha/2)^{1/n})$, and when $\hat{P} = 1$, the confidence interval is $((\alpha/2)^{1/n}, 1)$. For a more detailed explanation see Ott and Longnecker (2009).

In the previous example if we observe no red light violations in ten signal cycles, then the 95% confidence interval for the proportion of cycles having red light offenders is $(0, .3085)$. This is a much more credible answer than the 0 ± 0 obtained by using the usual formula when we have ten observations.

Our second caveat is a technical one. It is known that the CLT is not appropriate for some combinations of p and n. There are both lucky and unlucky combinations of these parameters, as described in Brown et al. (2001). For confidence intervals, we can do better than require $n\hat{P}$ and $n(1-\hat{P})$ to be at least 5. A more stable confidence interval is obtained by using the Agresti-Coull (1998) formula. For 95% confidence intervals, this is obtained by adding two successes and two failures to the data. Thus, the center of the Agresti-Coull confidence interval is

$$\hat{P}_{AC} = \frac{\sum_{i=1}^{n} X_i + 2}{n+4}. \tag{9.4}$$

We let

$$SE(\hat{P}_{AC}) = \frac{\sqrt{\hat{P}_{AC}(1-\hat{P}_{AC})}}{\sqrt{n+4}} \tag{9.5}$$

and the Agresti-Coull 95% confidence interval is

$$\hat{P}_{AC} \pm z_{1-\alpha/2} SE(\hat{P}_{AC}) \tag{9.6}$$

provided that the sample size is at least ten. Again, this is a general rule of thumb in statistics. For the red light example the 95% Agresti–Coull interval is (0, .33). This is reasonably close to the adjustment for the ordinary confidence interval. For reasons of scholarship, it should be noted that the Agresti–Coull interval is closely related to Wilson's (1927) interval.

When the sample size is small—say, less than 10—the authors advise using confidence intervals based upon the binomial distribution. Suppose that we wish to test the hypothesis that

$$H_o : p = p_0$$

$$H_a : p \neq p_0.$$

The sum $n\hat{P} = \sum_{i=1}^{n} X_i$ has a binomial distribution with parameters (n, p_0). Then, the exact p value for the test of hypothesis based upon a binomial distribution is

$$\sum_{\{|x - np_0| \geq |n\hat{p} - np_0|\}} \binom{n}{x} p^x (1-p)^{n-x}. \tag{9.7}$$

Confidence intervals can be calculated based upon the binomial approximation, but the allowable values of α are limited due to the small number of nonzero terms for the pdf of the binomial distribution. JMP does one-sided exact binomial tests for data that are recorded as zero or one and as nominal or ordinal variables. For further information see "Options for Categorical Variables" in the JMP manual. These calculations can be awkward if the sample size is large, and for sample sizes of ten and larger, the Agresti-Coull (1998) test and corresponding confidence intervals should suffice.

9.3 TESTS AND CONFIDENCE INTERVALS FOR TWO PROPORTIONS

Often we will want to compare the success rate of two traffic strategies. For example, we may want to know which of two signs are more effective at keeping traffic in the right lane of a two-lane road. One sign may say, "slower traffic keep right," and another sign may read, "left lane for passing only." In other instances, we may wish to know if having wider highway shoulders results in a smaller accident rate. In order to answer these questions, we generalize the approaches given in the last section. In this way we can test and obtain confidence intervals for the difference of two proportions.

Because we will want to know something about the success rate of the two procedures, we assume that we have independent observations from each of the two populations. Let $X_{1,1}, \ldots, X_{1,n_1}$ be n_1 independent Bernoulli observations from the first population with mean p_1. Also, let $X_{2,1}, \ldots, X_{2,n_2}$ be n_2 independent Bernoulli observations from the second population with mean p_2. Let $\hat{P}_1 = \sum_{i=1}^{n} X_{1i} / n_1$ and $\hat{P}_2 = \sum_{i=1}^{n} X_{2i} / n_2$ be the sample proportions from the two populations. Based upon the logic that we have used throughout this chapter, we will test the null hypothesis that $p_1 - p_2 = 0$ vs. the alternative hypothesis that $p_1 - p_2 \neq 0$ based upon how close the estimated difference is relative to its standard error. The estimated difference $\hat{P}_1 - \hat{P}_2$ has an estimated standard error of

$$SE(\hat{P}_1 - \hat{P}_2) = \sqrt{\left(\frac{\hat{P}_1(1 - \hat{P}_1)}{n_1} + \frac{\hat{P}_2(1 - \hat{P}_2)}{n_2} \right)}. \qquad (9.8)$$

Thus, we test the hypotheses:

$$H_0 : p_1 = p_2$$

$$H_a : p_1 \neq p_2$$

using a generalization of the single population approach. We use the test statistic

$$\frac{\left| \hat{P}_1 - \hat{P}_2 \right|}{SE(\hat{P}_1 - \hat{P}_2)} \qquad (9.9)$$

and reject H_0 if it is greater than $z_{1-\alpha/2}$. Otherwise, we do not reject H_0.

As before, we can also use a confidence interval to test the same hypothesis, and once again, it provides more information than the hypothesis test. The corresponding confidence interval for the difference of proportions $p_1 - p_2$ is

$$\hat{P}_1 - \hat{P}_2 \pm z_{1-\alpha/2} SE(\hat{P}_1 - \hat{P}_2). \qquad (9.10)$$

Just as in the case of a single proportion, the CLT may be inappropriate for certain values of p and n. As a result, the assumption that the test statistic is approximately normal may be inadequate. For the same reasons as previously stated, we assume that at least five successes and five failures from each population are mandatory for the normal approximation to have a reasonable chance of developing reasonable, and functional, confidence intervals.

Similar to the aforementioned example, we will present an improved alternative from Agresti and Caffo (2000). This formulation simply adds one success and one failure to the samples from each population, but otherwise proceeds in a similar manner. Again we add four observations in total. That is, each population has one success and one failure added. Let

$$\hat{P}_{AC,1} = \frac{\sum_{i=1}^{n} X_{1i} + 1}{n_1 + 2} \tag{9.11}$$

and

$$\hat{P}_{AC,2} = \frac{\sum_{i=1}^{n} X_{2i} + 1}{n_2 + 2}. \tag{9.12}$$

Then our new confidence interval is

$$\hat{P}_{AC,1} - \hat{P}_{AC,1} \pm z_{1-\alpha/2} \sqrt{\left(\frac{\hat{P}_{AC,1}(1-\hat{P}_{AC,1})}{n_1 + 2} + \frac{\hat{P}_{AC,2}(1-\hat{P}_{AC,2})}{n_2 + 2} \right)}. \tag{9.13}$$

It works much more reliably than the traditional confidence interval based upon a normal approximation.

9.4 CHI-SQUARE TESTS CONCERNING MORE THAN TWO POPULATION PROPORTIONS

Categorical data with multiple categories are often encountered in transportation. For example, incident types in the San Antonio, Texas DOT district are classified as major accident, minor accident, debris, and stalled vehicle. Crashes may be classified according to their severity level: fatal, incapacitating injury, nonincapacitating injury, minor injury, and property damage only. Moreover, vehicles may be classified according to their access maneuvers at HOV lanes and the use of the turn signals. As a first step, categorical data with more than two categories are usually summarized in a frequency table where each column corresponds to a particular category and the cells contain the frequency counts of observations.

9.4.1 Chi-Square Test for Univariate Categorical Data

Univariate categorical data can be summarized in a one-way frequency table. As a case in point, researchers randomly sampled 866 incidents

TABLE 9.1 One-Way Frequency Table for Randomly Selected Incidents in San Antonio, 2004 ($n = 866$)

	Type of Incident			
	Major Accident	Minor Accident	Stalled Vehicle	Debris on Road
Observed count	287	221	295	63

from the records of all incidents in San Antonio in 2004, and classified each according to type of incident, as shown Table 9.1.

The proportion of each incident type is computed to be 33% (= 287/866 × 100) for major accidents, 26% for minor accidents, 34% for stalled vehicle, and 7% for debris. Note that those percentages represent the sample proportions based on the observed data.

Sometimes transportation professionals would like to make inferences about categorical data, such as shown in Table 9.1. Let us denote the true proportions in each category of type of incident as p_{Major}, p_{Minor}, $p_{Stalled}$, and p_{Debris} for major accidents, minor accidents, stalled vehicle, and debris, respectively. Assume that there is a hypothesized value for each category denoted as p^0_{Major}, p^0_{Minor}, $p^0_{Stalled}$, and p p^0_{Debris}. Note that often the null hypothesis is that all proportions are the same (e.g., there is no effect of treatment). The null hypothesis and the alternative hypothesis of interest are stated as follows:

H_0: $p_{Major} = p^0_{Major}$, $p_{Minor} = p^0_{Minor}$, $p_{Stalled} = p^0_{Stalled}$, $p_{Debris} = p^0_{Debris}$.

H_a: At least one of the true category proportions is different from the corresponding hypothesized value.

The chi-square test can be used to evaluate the above hypotheses. This test is often referred to as a goodness-of-fit test (see, e.g., Agresti, 1990). The chi-square test statistic, denoted by χ^2, measures the discrepancy between the observed cell counts and the expected cell counts where the latter are given as the hypothesized category proportions multiplied by the sample size n. For example, the expected cell count for the category of major accidents is np^0_{Major}; for minor accidents, np^0_{Minor}; and so on. In general, when there are K categories, the chi-square statistic χ^2 can be defined as

$$\chi^2 = \sum_{k=1}^{K} \frac{(Y_k - np^0_k)^2}{np^0_k} \tag{9.14}$$

where Y_k denotes the observed cell count for the kth category and p_k^0 denotes the hypothesized value for the corresponding category. A large value of X^2 indicates that there is a significant discrepancy between the observed cell counts and the expected cell counts; thus, this information suggests a rejection of the null hypothesis. Yet, this begs the question: How do we define the term *large*? That is, how large should X^2 be in order to reject the null hypothesis? In practice, this can be determined based upon the corresponding p value. For example, a small p value (e.g., less than 0.05) corresponds to a large X^2 value. When the null hypothesis is true and the sample size is large enough so that $np_k^0 \geq 5$ for each category, the distribution of the statistic X^2 can be approximated by a chi-square distribution with degrees of freedom $df = K-1$ denoted by X_{K-1}^2. Therefore, the null hypothesis will be rejected if the p value associated with X^2 — computed based on the chi-square distribution with $df = K-1$—is less than the predetermined significance level, α.

Example 9.1

Assume that the data in Table 9.1 represent a random sample of the incidents that occurred in San Antonio in 2004. Imagine that you have been asked to determine if the four incident types occur with equal probability, and you have decided to conduct the test at $\alpha = 0.05$. In this situation the null hypothesis is $p_{Major} = p_{Minor} = p_{Stalled} = p_{Debris} = 0.25$, and the alternative hypothesis is that at least one of the true category proportions is not 0.25. The expected cell count for each category is computed as

Expected count $= n \times 0.25 = 866 \times 0.25 = 216.5$.

and the observed value of X^2 is

$$X^2 = \sum_{k=1}^{4} \frac{(Y_k - np_k^0)^2}{np_k^0}$$

$$= \frac{(287 - 216.5)^2}{216.5} + \frac{(221 - 216.5)^2}{216.5} + \frac{(295 - 216.5)^2}{216.5} + \frac{(63 - 216.5)^2}{216.5}$$

$$= 160.3464.$$

Because the expected cell counts (here 216.5) are all greater than 5, the p value can be approximated by the tail probability to the right of the observed

Level	Test Probabilities	
	Estimated Probability	Hypothesized Probability
Major accident	0.33141	0.25000
Minor accident	0.25520	0.25000
Stalled vehicle	0.34065	0.25000
Debris	0.07275	0.25000

Test	Chi-Square	DF	Probability > Chi-Square
Pearson	160.3464	3	<.0001

FIGURE 9.1   The JMP output for the chi-square test.

χ^2 value for the chi-square distribution with $df = 3$ (i.e., $\chi^2 \sim \chi_3^2$). This is represented as follows:

$$p \text{ value} = P(\chi^2 \geq 160.3464) < 0.0001.$$

In practice, the chi-square test is conducted by using statistical packages, such as JMP. By looking at the terms in the sum, we can make educated guesses about which classes cause the null hypothesis to be rejected. For example, the sum corresponding to minor accidents is

$$\frac{(221 - 216.5)^2}{216.5} \approx .09.$$

On the other hand, the term corresponding to debris is

$$\frac{(63 - 216.5)^2}{216.5} \approx 109.$$

It is very likely that debris, and not minor accidents, is contributing to rejecting the null hypothesis. Tests of proportions can be done as a more precise way to assess the cells causing the difference. For most applications, at least in the authors' experience, a check based upon the contributing terms to the χ^2 will suffice.

Figure 9.1 shows the JMP output for the chi-square test. Because the p value is less than α (= 0.05), H_0 is rejected. It can be concluded that there is strong evidence in the data that not all of the four incident types occur with equal probability.

9.4.2 Tests for Independence of Two Categorical Variables

When there are two categorical variables, the data can be summarized in a contingency table. A contingency table contains the frequency counts of observations classified according to two variables. The rows indicate

TABLE 9.2   Contingency Table of Access Maneuvers by Use of Turn Signal

Count	No	Yes	Row Marginal Total
Abort	29	26	55
In	2,249	2,236	4,485
Out	527	1,356	1,883
Passing	345	197	542
Column marginal total	3,150	3,815	6,965

different values of one variable, and the columns indicate the values of another variable. A contingency table having I rows and J columns is often referred to as an $I \times J$ table.

As part of a TxDOT-sponsored research project to develop guidance materials on intermediate access to a buffer-separated toll lane, researchers recorded whether a driver used the turn signal during the maneuver (Fitzpatrick et al., 2007). A contingency table for access maneuver at an HOV lane (abort, in, out, or passing) and use of turn signal (yes, no) is shown in Table 9.2. This is a 4×2 contingency table. Table 9.2 also shows the marginal totals for the columns and the rows, which are obtained by adding the cell counts in each column and in each row, respectively. The grand total, the sum of all observed cell counts in the table, is also given as 6,965.

A common question associated with categorical data of the sort shown in Table 9.2 is whether the two variables are independent. In this case, this question specifically asks whether maneuver type and use of turn signal are independent. By law, drivers must use a turn signal when conducting any of the maneuvers shown in Table 9.2. Clearly, some do not, and the transportation engineer might be interested in whether some drivers performing particular types of maneuvers are more likely not to use a turn signal indication than other types of maneuvers. In general, when there are two categorical variables and we wish to test whether there is any association between them, the null and alternative hypotheses can be stated as follows:

H_0: The two variables are independent.

H_a: The two variables are not independent.

Recall that the variables are statistically independent if all joint probabilities equal the product of their marginal probabilities. That is,

$$p_{ij} = p_i p_j \quad for \quad i = 1, \cdots, I, \; j = 1, \cdots, J$$

where p_{ij} denotes the joint probability that the response falls in the cell in the ith row and the jth column. Furthermore, p_i and p_j represent the marginal probability of the ith row (obtained by summing the joint probabilities in the ith row across the J columns) and the marginal probability of the jth column (obtained by summing the joint probabilities in the jth column across the I rows), respectively. Thus, under the null hypothesis of independence, the expected cell count for the ith row and jth column (E_{ij}) can be computed as

$$E_{ij} = \text{Expected count for the } (i, j)\text{th cell} = np_i p_j$$

where n is the sample size. Because the true marginal probabilities are unknown, those probabilities are estimated by the sample marginal proportions. The estimated expected cell count ($\hat{E}_{ij}$) can be obtained as follows:

$$\hat{E}_{ij} = n \,(j\text{th column marginal total}/n) \times (i\text{th row marginal total}/n)$$

$$= (j\text{th column marginal total}) \times (i\text{th row marginal total})/n.$$

The chi-square test can again be used to evaluate the above hypotheses. The chi-square test statistic, χ^2, is formed as the sum of the squared differences between the observed cell count and the estimated expected cell count divided by the estimated expected cell count. In general, when there are IJ cells in the contingency table, the chi-square statistic χ^2 can be defined as

$$\chi^2 = \sum_{i=1}^{I} \sum_{j=1}^{J} \frac{\left(Y_{ij} - \hat{E}_{ij}\right)^2}{\hat{E}_{ij}} \tag{9.15}$$

in which $Y_{i,j}$ and $\hat{E}_{ij}$ denote the observed cell count and the estimated expected cell count for the (i,j)th category, respectively. A large value of χ^2 indicates that there is a significant discrepancy between the observed cell counts and the expected cell counts, which suggests rejection of the null hypothesis. When the null hypothesis is true and the sample size is large enough so that $\hat{E}_{ij} \geq 5$ for each cell, the distribution of the statistic X^2 can be approximated by a chi-square distribution with degrees of freedom $df = (I-1)(J-1)$. The null hypothesis will be rejected if the p value approximated by the tail probability to the right of the observed X^2 value, for the chi-square distribution with $df = (I-1)(J-1)$, is less than the predetermined significance level, α.

Example 9.2

Assume that researchers are interested in testing whether there is any association between vehicles' access maneuvers at HOV lanes (abort, in, out, or passing) and the use of turn signals (yes or no) based on the data in Table 9.2. The hypotheses to be tested are as follows:

H_0: Access maneuvers at HOV lanes and the use of turn signals are independent.

H_a: Access maneuvers at HOV lanes and the use of turn signals are not independent.

The expected cell counts are estimated by

$$\hat{E}_{ij} = (j\text{th column marginal total})(\text{row marginal total})/(\text{grand total})$$

for $i = 1,2,3,4$ and $j = 1,2$, and are provided in parentheses in Table 9.3.

TABLE 9.3   Contingency Table of Access Maneuvers and Use of Turn Signal

Maneuver Type	Use of Turn Signal		Row Marginal Total
	No	Yes	
Abort	29	26	55
	(25)	(30)	
In	2,249	2,236	4,485
	(2,028)	(2,457)	
Out	527	1,356	1,883
	(852)	(1,031)	
Passing	345	197	542
	(245)	(297)	
Column marginal total	3,150	3,815	6,965

The observed value of χ^2 is

$$\chi^2 = \sum_{i=1}^{4}\sum_{j=1}^{2}\frac{\left(Y_{ij}-\hat{E}_{ij}\right)^2}{\hat{E}_{ij}}$$

$$= \frac{(29-25)^2}{25}+\frac{(26-30)^2}{30}+\cdots+\frac{(197-297)^2}{297}$$

$$= 345.241.$$

Because the estimated expected cell counts are all greater than 5, the p value can be approximated by the tail probability to the right of the observed χ^2 value for the chi-square distribution with $df = (2-1)(4-1)= 3$ (i.e., $\chi^2 \sim \chi_3^2$) as follows:

$$p \text{ value} = P(X^2 \geq 345.241) < 0.0001.$$

In practice, the chi-square test is conducted by using a statistics package, such as the JMP. Figure 9.2 shows the JMP output for the chi-square test.

Because the p value is less than α (= 0.05), H_0 is rejected. It can be concluded that there is strong evidence in the data that access maneuvers at HOV lanes and the use of turn signals are not independent. Once again, the individual terms in the sum defining the χ^2 can be used to judge which cells do not support independence by comparing them among one another. For example, the top left cell of the table has a contribution of $(29-25)^2/25$, which is less than 1, and the top left cell is not displaying dependence. On the other hand, the third row of the table has contributions such as $(527-852)^2/852 \approx 124$ and shows a strong dependence for that cell.

Tests

Source	DF	−Log-Like	R-Square (U)
Model	3	177.6634	0.0370
Error	6,961	4,618.3121	
C. total	6,964	4,795.9755	
N	6,965		

Test	Chi-Square	Probability > Chi-Square
Pearson	345.241	< 0.0001

FIGURE 9.2 The JMP output for the chi-square test of independence.

TABLE 9.4   Contingency Table of Rainfall by Incident Type

	Incident Type				Row Marginal Total
	Major Accident	Minor Accident	Stalled Vehicle	Debris	
No rain	153	124	173	45	495
Rain	90	57	60	13	220
Column marginal total	243	181	233	58	715

9.4.3 Tests for Homogeneity of Two or More Populations

In many contingency tables one variable may be fixed and play the role of a group variable that indicates the population from which the samples are selected, and a subject in each sample may be classified by the remaining variable. Suppose that researchers are interested in determining if the proportions of the four different incident types—major accidents, minor accidents, stalled vehicle, and debris—are the same for rainy days and days without rain. That is, they are interested in whether rain (the fixed group variable) has an effect on the number of each type of incident expected on any given day. To test this, suppose they collected a random sample of 220 rainy days and a separate random sample of 495 days with no rain. The data are summarized in the 2×4 contingency table (Table 9.4).

Table 9.4 also shows the marginal totals for the columns and the rows, which are obtained by adding the cell counts in each column and in each row, respectively. Here, the two row marginal totals, 495 and 220, are fixed by the sampling design (e.g., the sample sizes for nonrainy days and rainy days were fixed by the experimenter) while the four column totals are random. The grand total, the sum of all observed cell counts in the table, is also given as 715.

As stated above, the hypothesis of interest here is whether the proportions of the four different incident types are the same for a population of rainy days and a population of days without rain. In general, when there are I populations from which separate random samples are obtained and each observation in the sample is classified according to the value of a categorical variable, the null and alternative hypotheses of interest can be stated as follows:

H_0: The true category proportions are the same across all of I populations.

H_a: The true category proportions are not the same for all of I populations; that is, they are different for at least one of the I populations.

Note that the null hypothesis represents homogeneity of populations with respect to the category proportions. The chi-square test statistic can again be used to test the above hypotheses. The expected cell count under the null hypothesis for the jth category and the ith population ($\hat{E}_{ij}$) can be estimated as follows:

$\hat{E}_{ij}$ = (sample proportion of the jth category) × (ith sample size)

= (jth column marginal total/grand total) × (row i marginal total)

= (jth column marginal total) × (row i marginal total)/(grand total).

As before, the chi-square test statistic, χ^2, is formed as the sum of the squared differences between the observed cell count and the estimated expected cell count divided by the estimated expected cell count. In general, when there are IJ cells in the contingency table, the chi-square statistic χ^2 can be defined as

$$\chi^2 = \sum_{i=1}^{I}\sum_{j=1}^{J} \frac{\left(Y_{ij} - \hat{E}_{ij}\right)^2}{\hat{E}_{ij}} \qquad (9.16)$$

where $Y_{i,j}$ and $\hat{E}_{ij}$ denote the observed cell count and the estimated expected cell count for the (i,j)th category, respectively. A large value of χ^2 indicates that there is a significant discrepancy between the observed cell counts and the expected cell counts; thus, this suggests rejection of the null hypothesis. When the null hypothesis is true and the sample size is large enough so that $\hat{E}_{ij} \geq 5$ for each cell, the distribution of the statistic χ^2 can be approximated by a chi-square distribution with degrees of freedom $df = (I-1)(J-1)$. The null hypothesis will be rejected if the p value approximated by the tail probability to the right of the observed χ^2 value for the chi-square distribution with $df = (I-1)(J-1)$ is less than the predetermined significance level, α.

Example 9.3

Suppose that researchers are interested in testing whether the proportions of the four different incident types—major accidents, minor accidents, stalled vehicle, and debris—are the same for rainy days and days without rain based on the data in Table 9.4. The hypotheses to be tested can be stated as follows:

H_0: The category proportions of incident type are the same for rainy days and days with no rain.

H_a: The category proportions of incident type are not the same for rainy days and days with no rain.

TABLE 9.5   Contingency Table of Rainfall by Incident Type with Estimated Expected Counts

Count	Major Accident	Minor Accident	Stalled Vehicle	Debris	Row Marginal Total
No rain	153	124	173	45	495
	(168)	(125)	(161)	(40)	
Rain	90	57	60	13	220
	(75)	(56)	(72)	(18)	
Column marginal total	243	181	233	58	715

The estimated expected cell counts computed by

$\hat{E}_{ij}$ = (jth column marginal total)(row i marginal total)/(grand total)

for $i = 1,2$ and $j = 1,2,3,4$, and are provided in parentheses in Table 9.5. The observed value of X^2 is

$$\chi^2 = \sum_{i=1}^{2}\sum_{j=1}^{4}\frac{\left(Y_{ij}-\hat{E}_{ij}\right)^2}{\hat{E}_{ij}}$$

$$= \frac{\left(153-168\right)^2}{168} + \frac{\left(124-125\right)^2}{125} + \cdots + \frac{\left(13-18\right)^2}{18}$$

$$= 9.181.$$

Because the estimated expected cell counts are all greater than 5, the p value can be approximated by the tail probability to the right of the observed χ^2 value for the chi-square distribution with $df = (2 - 1)(4 - 1) = 3$ (i.e., $\chi^2 \sim \chi_3^2$) as follows:

$$p \text{ value} = P(\chi^2 \geq 9.181) = 0.0270.$$

In practice, the chi-square test is conducted by using a statistics package such as JMP. Figure 9.3 shows the JMP output for the chi-square test.

Because the p value is less than α (= 0.05), H_0 is rejected. We can conclude that there is sufficient evidence in the data that the category proportions of incident type are not the same for rainy days and days with no rain. That is, the presence or absence of rain appears to affect the types of incidents that are observed. As before, individual contributions to the chi-square statistic can be used to gain an understanding of which cells contribute to the null hypothesis being rejected.

When the assumptions for the large sample tests are not met, i.e., some of the (estimated) expected cell counts are less than 5, the chi-square

Tests

Source	DF	−Log-Like	R-Square (U)
Model	3	4.62130	0.0050
Error	709	913.22117	
C. total	712	917.84247	
N	715		

Test	Chi-Square	Probability > Chi-Square
Pearson	9.181	0.0270

FIGURE 9.3 The JMP output for the chi-square test of homogeneity of two populations.

approximation may be invalid and exact tests (see, e.g., Agresti, 1990) may need to be employed. The reader is referred to Agresti (1990) for more in-depth discussion on inferences on various contingency tables.

9.5 THE CHI-SQUARE GOODNESS-OF-FIT TEST FOR CHECKING DISTRIBUTIONAL ASSUMPTIONS

The chi-square goodness-of-fit test can also be used to test whether the underlying distribution of the data follows a specific distribution (e.g., normal, exponential, etc.). The test is based upon the multinomial distribution.

Suppose that we obtain a set of random data and construct a histogram. We are interested in what the underlying distribution could be. The first step is to examine the histogram and hypothesize what would be a reasonable distribution type. Figure 9.4 shows a histogram of headway data obtained from a freeway. Given the fact that the headway decreases at a decreasing rate, it might be reasonable to hypothesize that the underlying distribution is exponential. We could overlay an exponential distribution on the histogram in order to visually compare the histogram and the density function.

Let E_i be the expected number of observations within the ith cell of the histogram based upon an assumed theoretical pdf. Assume that the null hypothesis is correct, and let n_i be the number of observations within the ith cell. Then the chi-square test statistic is

$$\chi^2 = \sum_{i=1}^{I} \frac{(n_i - E_i)^2}{E_i}. \tag{9.17}$$

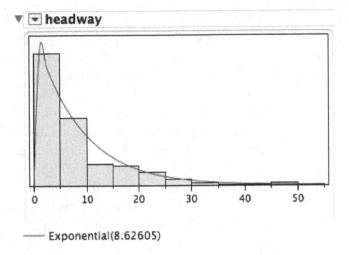

FIGURE 9.4 Histogram for headway data with an overlaid exponential distribution.

This statistic has an approximate chi-square distribution with $I-1$ degrees of freedom. We reject the null hypothesis if the value of χ^2 exceeds the $100(1-\alpha)$ percentile of a chi-square distribution with $I-1$ degrees of freedom.

As an example, consider the two hundred headway observations used in Figure 9.4. Suppose we use five equally spaced cells to construct our chi-square test. Note that at this time JMP does not easily perform a chi-square test for continuous distributions. Thus, other packages, such as MATLAB, may be used. Using the MATLAB "chitest" command we get $\chi^2 = 10.5419$, and for a chi-square with 4 degrees of freedom, the p value is .0322. The p value obtained from JMP using the Kolmogorov test is less than .01 for these data. The p value given by the chi-square test does not properly take into account the estimation of the unknown parameter in the exponential distribution. However, in large samples this should not be very important.

9.6 CONCLUSIONS

Categorical data are used regularly by transportation agencies and, as such, are often analyzed by engineers. This is because some transportation-related factors are, by definition, discrete. Examples of these could include (1) number of lanes on a roadway, (2) personal characteristics such as gender, (3) experiential characteristics such as whether someone has taken a traffic course, (4) specific treatment such as type of de-icer, and (5) number of cars in a household. In other situations the variables are discrete because that is how the information was collected. For example, it is rare to ask specific questions related to salary; rather, the survey typically asks responders to fill in their answer as part of a range. When the data are categorical, the transportation professionals often ask questions related to proportions of the categorical data.

The techniques used to analyze one or two proportions are very similar to those related to the inference of the estimated mean value of continuous variables. Both hypothesis tests and confidence intervals can be used for inference questions. As previously explained, the authors recommend using confidence intervals for inference. The results will be the same as for standard hypothesis testing; however, the authors feel the CIs provide more useful information. Note that when there were more than three proportions being compared, only hypotheses tests were presented.

This chapter also introduced a number of issues that are particular to categorical inference. For example, if no events are observed for a particular situation, the traditional CI calculations will give answers that are clearly inappropriate. However, with the revised CI this issue is handled, as the modified CIs are more accurate than the standard approaches. As before, it is imperative for the analyst to understand the limitations of the various approaches.

The concept of two-way contingency tables was also developed in this chapter. These are very common in transportation studies. For example, transportation agencies are often interested in whether a variable they have control over, such as a type of sign or roadway, has an effect on some type of driver characteristic. The approaches would then allow for various questions to be answered, such as whether the presence of education campaigns affects the number of people running through red-lights or texting while driving.

HOMEWORK PROBLEMS

1. In a study of red light offenses at an intersection, there were 180 light cycles. In 163 cycles, there were no red light violations, and in 17, there was at least one vehicle that ran a red light. Give point and interval estimates for the mean percent of red light violations.

2. A state department of transportation is interested in knowing whether older Americans can drive safely (compared to the general population) if their blood alcohol content (BAC) is .06%, or 75% of the state's statutory limit. Fifty volunteers are used: thirty participants

are under the age of forty-five, and twenty are over the age of sixty. Each participant is given alcoholic beverages at fifteen-minute intervals until his or her BAC is .06%. Three drivers from the under the age of forty-five group and five from the older group had driving mishaps. Provide confidence intervals for the mean difference of the proportion of drivers adversely effected by a BAC of .06%.

3. A transportation engineer is comparing two strategies for cutting down on red light violations. The first strategy uses warning signs of the upcoming traffic signal. The second uses sensors to keep signals green if a vehicle is calculated not to have enough time to easily stop for a red signal. The engineer records driver behavior at two matched intersections. Are the sensors successful at cutting down the percentage of red light runners?

	Stopped at Red	Continued on Red
Warning signs	304	9
Sensors	256	6

4. A work zone study is being conducted to see what variables most influence a driver's decision to decrease speed. The study uses volunteers from passive and active states. In a passive state, police are present at work zones but do not ticket speeders. In an active state, police are present at work zones and ticket speeders. Use a test of homogeneity to see if there are significant age differences between passive and active states in the study. If there are differences, what age categories are contributing to the differences?

Enforcement	A	B	C	D	E	F
Active	83	150	223	60	21	22
Passive	79	146	215	108	34	17

REFERENCES

Agresti, A. 1990. *Categorical data analysis.* New York: John Wiley & Sons.

Agresti, A., and B. Caffo. 2000. Simple and effective confidence intervals for proportions and difference of proportions result from adding two successes and two failures. *The American Statistician* 54(4):280–288.

Agresti, A., and B. A. Coull. 1998. Approximate is better than "exact" for interval estimation of binomial proportions. *The American Statistician* 52:119–26.

Brown, L. D., T. T. Cai, and A. DasGupta. 2001. Interval estimation for a binomial proportion. *Statistical Science* 16:101–17.

Fitzpatrick, K., M. A. Brewer, and E. S. Park. 2007. *Intermediate access to buffer-separated managed lanes*. Research Report 0-5547-1. College Station, TX: Texas Transportation Institute.

Ott, R., and M. Longnecker. 2009. *An introduction to statistical methods and data analysis*. 6th ed. Belmont, CA: Cengage/Brooks/Cole.

Wilson, E. B. 1927. Probable inference, the law of succession and statistical inference. *Journal of the American Statistical Association* 22:209–12.

CHAPTER 10

Linear Regression

10.1 INTRODUCTION

In previous chapters, we looked at modeling single transportation phenomena, such as vehicle speed, as a random variable. We illustrated how to graph empirical speed data, make hypotheses about the distribution of speeds, and then test the hypotheses using statistical techniques, such as the Kolmogorov–Smirnov (K–S) test. We also showed how we quantify the random variable by measures of central tendency (mean, mode, median) and measures of dispersion (variance and range). Additionally, we demonstrated how to test hypotheses about these measures by using confidence intervals (CIs), such as t-tests and F-tests. In many transportation applications, however, we are interested in the *relationship* between two variables. In particular, engineers are often asked what happens to one variable if another variable is changed; that is, they are asked to predict outcomes. In this situation we will need to know how to develop a statistical relationship between the two variables. For example, we might be interested in the relationship between two variables in several scenarios, such as (1) the service life of a pavement and average daily traffic (ADT), (2) visibility distance and sign font size, (3) vehicle speed and the posted speed limit, or (4) crash rate and speed limit.

As engineers we would like to make rationale trade-offs with a variable we "control" (e.g., pavement thickness, font size, speed limit) in order to meet some objective (e.g., increases in pavement life, sign visibility, safety). However, we need to have a model that relates the two variables, and we must to be able to test the model for accuracy. For example, a common assumption by novices is that by reducing the speed limit, vehicle speeds will automatically decrease. In this chapter it will be shown how a model that can be used to test this hypothesis can be developed and utilized. This chapter is focused on modeling a particular subset of all these types of relationships, or the situation where the relationship between two variables is *linear in the parameters*, that is, where the response of a particular variable is linear in nature (at least in the particular range of data under examination). We start this chapter using a straight-line example, and then proceed to more complicated examples.

10.2 SIMPLE LINEAR REGRESSION

If we believe that the relationship between two variables is linear, then the modeling process consists of collecting paired data on the two variables and fitting a straight line. Fitting a straight line is among the simplest and most powerful tools in statistical modeling and would be familiar, in its basic form, to any elementary school student. Statisticians call the fitting of a straight line simple linear regression, but perhaps a more appropriate description is "straightforward" rather than simple. We will also illustrate how to test the hypothesis that a relationship is linear.

Let us consider Figure 10.1, which is a scatter plot showing the average annual delay on the vertical axis and the number of principal arterial streets across major metropolitan areas on the horizontal axis. These

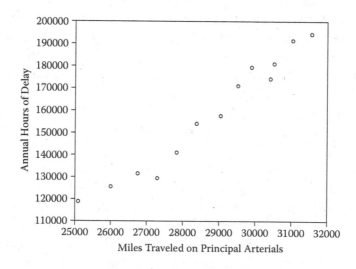

FIGURE 10.1 Bivariate fit of annual hours of delay by miles traveled on principal arterials.

data were obtained from a collection of cities used to compute the Texas Transportation Institute Congestion Index (Schrank and Lomax, 2007).

Based on a cursory examination of the graph, it may be concluded that as the daily vehicle miles traveled (in thousands of miles) on principal arterial streets increases, the number of annual hours (in thousands) of delay increases. There is a very important lesson behind this observation. As an engineer, you need to identify whether a relationship is causal before you can use your model for engineering purposes. For example, if you were to take a given city and build more major arterials, you would not expect an increase in delays. Rather, the above relationship exists because both delay and the number of major arterials are related to another variable: city population size. This does not mean that the model is inaccurate or that it cannot be used for prediction. For example, the number of arterial roadways may, in fact, be a good predictor of delay. If you wanted to estimate delay in a city, but only had information on the number of arterial streets, this might be an appropriate model. However, in this situation you will not be able to directly manipulate one variable (e.g., add more streets) and predict the outcome of the other.

> **Note:** If you wish to manipulate one variable and predict the outcome of another, you will need to establish that there is a causal relationship. Sometimes this is very easy to do; for example, it is easy to establish that vehicle momentum is directly related to crash severity, all else being equal. However, in other situations, this is not so easy and many false results can be obtained. Situations like the example above, in which there is a hidden variable (which causes the relationships in the other two), are known as false cause fallacies.
>
> **There is *no* statistical test that can be used to determine causation. The only tools that are typically available in these situations are scientific knowledge and engineering judgment.**

In summary, the methods presented in this chapter should be considered as descriptive or predictive, but they do not imply underlying causes.

10.2.1 Correlation Coefficient

The first descriptive measure we discuss is correlation, or the Pearson correlation coefficient. Suppose that we have paired data (X_i, Y_i), $i = 1, \dots, n$. The Pearson correlation coefficient, r, introduced in Chapter 3 gives a measure of linearity between the two variables in the data. It is also called the sample correlation coefficient. As presented in Chapter 3, the sample correlation coefficient is defined by the equation

$$r = \frac{\dfrac{1}{n}\sum_{i=1}^{n}(X_i - \bar{X})(Y_i - \bar{Y})}{\sqrt{\dfrac{1}{n}\sum_{i=1}^{n}(X_i - \bar{X})^2}\sqrt{\dfrac{1}{n}\sum_{i=1}^{n}(Y_i - \bar{Y})^2}}. \qquad (10.1)$$

For the data in Figure 10.1, $r = .981$.

Correlations are useful for quickly measuring the degree of linearity among variables. They are particularly useful for scanning data sets with many variables to find those that are linearly related. In other words, if you had a data set with fifty variables, you would need $50 \times 49/2 = 1225$ scatter plots to visually examine all the relationships. Thus, even though good statistical practice is to plot data before they are analyzed, plotting is not always practical.

Note: A value of $r = 0$ indicates the relationship between the two variables is not linear. However, this does not imply that there is no relationship between the variables; rather, it only indicates that if there is a relationship, it is not linear. The reader may wish to verify that the sample correlation coefficient between the data (X_i, Y_i), where $Y_i = X_i^2$ and $X_i = \pm(i/10)$, $i = 1,...10$, has $r = 0$. Clearly, there is a relationship between Y and X. Accordingly, we need to recognize that we may overlook nonlinear relationships when we scan through high-dimensional data using correlations.

There is a corresponding population quantity to the sample correlation coefficient defined in Equation 10.1, which measures the strength of the linear relationship between two variables in the entire population. It is called the population correlation coefficient and denoted by ρ. In fact, the correlation (ρ_{XY}) introduced in Chapter 4 corresponds to the population correlation coefficient (ρ).

JMP gives p values for correlation coefficients when the hypotheses of interest are $H_0 : \rho = 0$ vs. $H_a : \rho \neq 0$. In JMP, this feature is found in the multivariate menu. When both X and Y can be viewed as random, the standard error for r is approximately

$$s.e.(r) = \sqrt{\frac{1-r^2}{n-2}}. \qquad (10.2)$$

This standard error can be used to get approximate confidence intervals for a population correlation, ρ. An approximate confidence interval is

$$r \pm z_{1-\alpha/2} s.e.(r). \qquad (10.3)$$

Please note that the confidence interval should be truncated when necessary to avoid extending beyond ± 1. We would like to emphasize that JMP uses a more complicated and accurate confidence interval formula, which is based on a transformation of the estimated correlation. Those interested in further information on this approach may find a detailed description in the JMP manual.

10.2.2 Fitting a Straight Line

Often, knowing that two variables are linearly related is the first step in a study. The next step is to identify the formula of the straight line. Specifically, we want to discover the best straight lines out of the infinite number that are possible (or more realistically, from all the lines that might reasonably describe the linear relationship). To further complicate the issue, there are many straight lines that can be fitted when at least one of the variables has a measurement error.

As an example, Figure 10.2 shows the relationship between the annual hours of delay and the number of principal arterial streets in Beaumont, Texas. The solid line represents the typical least-squares line (we will describe how the least-squares line was obtained later), and the dotted line represents the fit when labeling of the X-Y pairs is reversed (e.g., instead of regressing y on x, we regress x on y). We could have drawn a line freehanded, and that would probably be different from both of these. There are other lines that are used as well, and most of them lie between those two lines.

But, which of these lines should we use as our model? To answer this question, we will need to discuss the errors related to the various models.

The basic linear model has two components. The first component is the underlying "true" relationship, which is the straight-line model for the mean of Y: $E(Y) = a + bX$. The second component is the error that incorporates all deviations from a straight line. These deviations may be due to measurement error or to the fact that there are other unknown variables

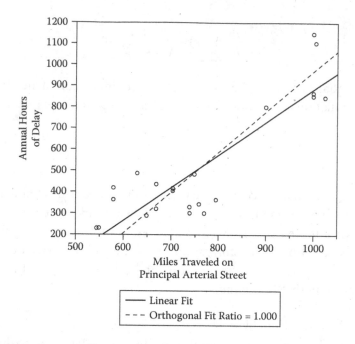

FIGURE 10.2   Bivariate fit of annual hours of delay by miles traveled on principal arterial street.

that affect the relationships. For most statistical straight-line fitting techniques, these errors are assumed to be normally distributed. We will show how this assumption may be checked later in this chapter.

In Figure 10.2, the underlying relationship between annual hours of delay and the miles driven on principal arterial streets is not (perfectly) linear. Part of the error is systematic, or what statisticians call *bias*, and it represents *error* in the model. The second component of the error is random measurement error, and it behaves randomly. Hours of delay are estimated with error, and this is part of the reason that the observations do not fall exactly on the least-squares line. Thus, the model that underlies simple linear regression is

$$Y = a + bX + \varepsilon. \qquad (10.4)$$

The X variable is commonly referred to as the independent variable or predictor. The term *predictor* is a more appropriate name because sets of independent variables can be correlated or dependent, leading to awkward terminology. In this textbook we use the term predictor. The Y variable is known as the dependent variable. The a and b are the intercept and slope parameters that we wish to estimate. The ε is the error in the model. We assume that we have observations (X_i, Y_i), $i = 1,...,n$, where $Y_i = a + bX_i + \varepsilon_i$. We also *assume* that the errors are normally distributed with mean zero and constant variance, which implies that the systematic offsets or biases are negligible. The assumption of normality is typically used, although it can often be replaced by an assumption that the error distribution produces few outliers. When there are many outliers, the usual confidence intervals for the parameters are often wide, and it is difficult to estimate the range of the true slope and intercept. In these cases, robust fits to the data are recommended as described in Mosteller and Tukey (1977).

It should be noted that the residuals (e.g., the difference between the observed value of Y and the predicted value) should be saved and plotted. A box plot will be used to check for outliers, and a q-q plot will be used to assess normality. These plots will be covered in greater detail later in the chapter.

10.2.2.1 Estimating Slope and Intercept

The least-squares estimators $\hat{a}$ and $\hat{b}$ are obtained by a minimization of the errors between the observations and a straight line as defined in the expression

$$\sum_{i=1}^{n}\left(Y_i - a - bX_i\right)^2$$

with respect to a and b. The formulas for estimating the slope (b) and the y intercept (a) are as follows:

$$\hat{b} = \frac{\sum_{i=1}^{n}\left(X_i - \bar{X}\right)\left(Y_i - \bar{Y}\right)}{\sum_{i=1}^{n}\left(X_i - \bar{X}\right)^2} \tag{10.5}$$

and

$$\hat{a} = \bar{Y} - \hat{b}\bar{X}. \tag{10.6}$$

The above formulas are known in the statistical literature as best linear unbiased estimates (BLUEs). While not covered here, they are generally taken as representing the best straight line out of all that are possible. In this situation, the best is the resulting straight line that has the lowest sum of squared errors out of all possible lines, and the smallest standard error possible. For this reason, the BLUE model shown above is the best model for most engineering applications, and is the one readers will most likely encounter in their professional careers.

Note: These formulas are so popular that they can be found in most spreadsheet packages (such as Excel) and scientific calculators. While there may be a temptation to use these formulas with a calculator, it should be resisted. The formulas are theoretically correct, but they are prone to round-off errors. Many packages, such as JMP, use numerically stable algorithms to calculate these same quantities.

10.2.2.2 Inferences on Slope and Intercept

In essence, the regression equation allows us to estimate a measure of central tendency, in this case, the mean for a given value of input x. Intuitively it would be useful to test hypotheses related to intercept, slope, and the line. Standard tests have been developed over the years to address these issues, and these are often output by most computer packages. Similar to the univariate case, we need an estimate of the variance. The first formula of interest is for the estimated variance for the residuals:

$$\hat{\sigma}^2 = \frac{\sum_{i=1}^{n}\left(Y_i - \hat{a} - \hat{b}X_i\right)^2}{n-2}. \tag{10.7}$$

It is evident that the estimated variance is very similar to that of the univariate case. The only difference in the numerator is that instead of using the estimated mean, the estimated mean for a given value of x is used. The denominator is $n-2$ rather than $n-1$ to account for the fact that both the slope and intercept were estimated (e.g., in statistical parlance we lose a degree of freedom).

The estimated standard error for the estimated slope is

$$\hat{\sigma}_{\hat{b}} = \frac{\hat{\sigma}}{\sqrt{\sum_{i=1}^{n}(X_i - \bar{X})^2}}. \tag{10.8}$$

The estimated standard error for the estimated intercept is

$$\hat{\sigma}_{\hat{a}} = \hat{\sigma} \sqrt{\frac{1}{n} + \frac{\bar{X}^2}{\sum_{i=1}^{n}(X_i - \bar{X})^2}} \qquad (10.9)$$

The degrees of freedom associated with these standard error estimates are $n - 2$. Note that the denominator in Equation 10.9 has a term related to difference in the values of the x predictor minus the average value of the x predictor. In essence, obtaining a relatively wide range of x values will allow for stronger statistical testing than having a small range, all else being equal. Caution is needed not to choose predictor values too far apart because the model may not be valid over too wide a range of the predictor.

As before, we can calculate confidence intervals and use these to test hypotheses. The CIs for the estimated slope (b) and intercept (a) are

$$\hat{b} \pm t(1 - \alpha/2, n-2)\hat{\sigma}_{\hat{b}} \qquad (10.10)$$

and

$$\hat{a} \pm t(1 - \alpha/2, n-2)\hat{\sigma}_{\hat{a}} \qquad (10.11)$$

respectively. Just as in ANOVA (see Chapter 8), computer programs, such as JMP, provide an ANOVA table with their regression output. The F-test in the output tests whether or not the slope equals zero, and is the most commonly utilized test related to the slope. If we cannot tell whether the slope is statistically different from zero, then it is impossible to argue that there is a linear relationship between the variables. The null hypothesis underlying this test is that *the population slope equals zero*, and the alternative hypothesis is that *the population slope is not equal to zero*. Alternatively, if the confidence interval for slope is calculated, then the equivalent hypothesis test can be done by checking whether zero lies in the confidence interval or not. The important regression output from JMP for the plot shown in Figure 10.2 is shown in Figure 10.3.

In Figure 10.3 it is clear that the mean square for error, that is, our variance estimate ($\hat{\sigma}^2$), is 20,111. It is important to remember that statistical packages will not provide the units, and consequently, you must keep

Linear Fit

Annual Hours of Delay = −664.6465 + 1.5443783*Miles Traveled on Principal Arterial Street

Summary of Fit

RSquare	0.759012
RSquare Adj	0.746963
Root Mean Square Error	141.8125
Mean of Response	512.5909
Observations (or Sum Wgts)	22

Analysis of Variance

Source	DF	Sum of Squares	Mean Square	F Ratio
Model	1	1266815.6	1266816	62.9918
Error	20	402215.7	20111	Prob > F
C. Total	21	1669031.3		<.0001*

Parameter Estimates

Term	Estimate	Std Error	t Ratio	Prob>|t|
Intercept	−664.6465	151.3777	−4.39	0.0003*
Miles Traveled on Principal Arterial Street	1.5443783	0.194586	7.94	<.0001*

FIGURE 10.3   Regression statistics for Figure 10.2.

track of them. The *F*-ratio in the ANOVA table is the square of the *t*-ratio for the slope estimate in the table. Accounting for rounding error is demonstrated in the following equation:

$$t_b^2 = (7.94)^2 = 63.0436 \approx 62.9918.$$

Using only two decimal places in our calculation causes a slight offset in our value for t_b^2. The p values for the *F*-test and the *t*-test for the slope will always be identical. Based on the regression output in Figure 10.3, we could conclude that both the intercept and the slope are significantly different compared to zero.

> **Note:** When the intercept is shown to be not statistically different than zero, some textbooks advocate "forcing" the line to run through zero. This tactic only estimates the value of the slope. In practice, it is usually best to estimate the intercept unless there is strong reason to believe the line runs through the origin.

10.2.3 Prediction

A common reason for fitting a straight line is to predict the value of the dependent variable Y from the given value of the predictor variable X. For example, we may wish to predict travel times using measured volume from inductance loop detectors as a predictor. For prediction individual parameters are less important; instead, it is pertinent to know how the estimated parameters are used together to make a prediction. The predictor that we will use is

$$\hat{Y} = \hat{a} + \hat{b}X. \tag{10.12}$$

The standard error for the prediction is

$$\hat{\sigma}_{\hat{Y}} = \hat{\sigma} \sqrt{1 + \frac{1}{n} + \frac{(X - \bar{X})^2}{\sum_{i=1}^{n}(X_i - \bar{X})^2}}. \tag{10.13}$$

The standard error for the line used in prediction is

$$\hat{\sigma}_{\hat{a}+\hat{b}X} = \hat{\sigma} \sqrt{\frac{1}{n} + \frac{(X - \bar{X})^2}{\sum_{i=1}^{n}(X_i - \bar{X})^2}}. \tag{10.14}$$

The difference between the two formulas in Equations 10.13 and 10.14 is the scalar 1 inside the square root sign. The confidence interval for the line (e.g., the average Y value) at point x is

$$\hat{a} + \hat{b}x \pm t_{n-2,1-\alpha/2} \hat{\sigma}_{\hat{a}+\hat{b}x}. \tag{10.15}$$

The interval estimate for a new predicted Y (future observation) at x, which is called the *prediction interval*, is given as

$$\hat{a} + \hat{b}x \pm t_{n-2,1-\alpha/2} \hat{\sigma}_{\hat{Y}}. \tag{10.16}$$

That addition of the scalar 1 in Equation 10.13 is needed to expand the confidence interval enough so that the resulting prediction interval is sufficiently wide enough to capture the unobserved Y (new points on a scatter plot) at the desired confidence level. If the errors in the Y are nonnormal—with heavy tails so that the data are prone to outliers—then the approach described above is not appropriate. In this situation the authors advise the reader to obtain statistical assistance from a professional so that appropriate prediction intervals can be constructed. The prediction interval given in Equation 10.16 can be plotted in JMP by selecting the appropriate menu choice after fitting a line. The resulting plot will resemble Figure 10.4.

In Figure 10.4 the shaded region shows the 95% confidence interval for the fitted line at each x. Note that the CI for the simple regression case is directly analogous to the CI for the single variable situation. Basically, if we were to repeat the same experiment one hundred times, we would expect the CI to contain the true mean ninety-five times.

In contrast, the wider intervals, which are indicated by the dotted line, show the 95% prediction intervals for the predicted Y at each x. Notice how the prediction intervals capture the points on the graph (Ys) and the confidence intervals capture the fitted line with a tighter band. This demonstrates the fact that prediction intervals need to be wider than confidence

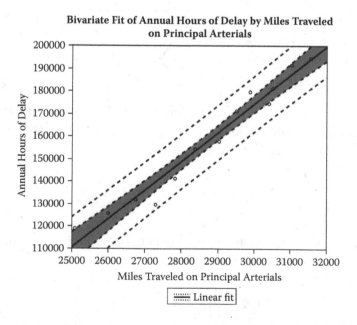

FIGURE 10.4   Prediction intervals, confidence intervals, and the fitted line.

intervals in order to capture observations. In summary, we use the CI to measure how well we can predict the average mean value for a given x value and the prediction interval (PI) to measure the range of observations we might expect for a given x value.

10.3 TRANSFORMATIONS

Often two variables are related in such a way that a transformation of at least one of the variables leads to a linear relationship between the two. When two variables have a monotonic relationship (such as $y = e^x$ for nonnegative variables), $y = x^2$ transformations, such as the log or square root, can be applied to create a linear relationship. Thus, engineers should consider transformations to their data when scatter plots indicate that x and y do not have a linear relationship. As an example, the scatter plot in Figure 10.5 shows annual hours of delay vs. daily freeway miles traveled in thousands for Akron, Ohio. Two curves are plotted. One is a straight-line fit (dashed line) and the other is a $\ln(y)$ to $\ln(x)$ straight-line fit (solid curve). The latter was created by first transforming the x and y data using a natural logarithmic function. The transformed data were then used in a simple linear regression analysis as described in this chapter. The resulting $\ln(y)$-$\ln(x)$ regression line was subsequently transformed back to its original units and plotted in Figure 10.2 as a solid line. An inspection shows

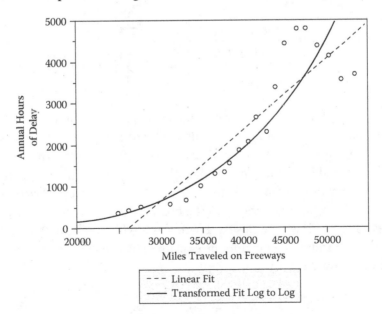

FIGURE 10.5 Bivariate fit of annual hours of delay by miles traveled on freeways.

that this latter curve appears to fit the data better, and thus the researcher may wish to start her analysis assuming a logarithmic relationship.

One downside of transformation is that the resulting estimates and confidence intervals obtained from the transformed data are for transformed variables, such as log-transformed passenger miles and not actual passenger miles. It is usually important for the transportation professional to transform answers back to the original units. The curve in Figure 10.5 illustrates a linear regression line that has been transformed back to its original units. An excellent book about this important topic is by Carroll and Rupert (1988).

10.4 UNDERSTANDING AND CALCULATING R^2

Even though we are using the BLUE estimator, a reasonable question arises: How good is our model? A commonly used measure of model fit in linear regression is R^2, which is called the coefficient of determination. For the simple linear regression, R^2 is equal to the squared Pearson sample correlation: $R^2 = r^2$. It can be calculated two ways in simple linear regression. The first way is to square the sample correlation coefficient. The second is to use the equation

$$R^2 = \frac{\text{Model Sum of Squares}}{\text{Total Sum of Squares}} = \frac{\sum_{i=1}^{n}(\hat{Y}_i - \bar{Y})^2}{\sum_{i=1}^{n}(Y_i - \bar{Y})^2}. \qquad (10.17)$$

This latter approach is more general, as it applies beyond simple linear regression, and therefore is recommended by the authors. It can also be used in multiple linear regression, as will be discussed later in this chapter. From the above equation, it is clear that R^2 measures the proportion of the total variation attributable to regression. The value of R^2 lies between 0 and 1. When most variability in the observed y can be explained by the regression line, R^2 will be close to 1, whereas it will be close to 0 when most variation is random and cannot be explained by the regression line.

In the example given in Figure 10.3,

$$R^2 = \frac{1266815.6}{1669031.3} = .759.$$

There is no best overall methodology to use to compute R^2, as it really depends on which values are easiest to obtain. Given that the vast majority of the calculations will be done by the statistical package, deliberating over which method to use is quite unnecessary.

Of more importance is whether the value of R^2 should be considered acceptable. Unfortunately, there is no set answer. In some areas of transportation engineering, such as economic impact studies, $R^2 \geq .6$ is considered good. In other areas, such as calibrating double-loop speed measurements to radar gun speed measurements, $R^2 < .9$ may be considered poor. In our example, $R^2 = .759012$, and the reader should check to be sure that he or she can calculate r using this value for R^2.

Finally, R^2 is only a crude measure of linearity. Consider the (X,Y) data pairs $(0,3),(0,3),(1,3)$. With this given data set, the R^2 for the straight-line fit to these data equals zero. Now consider adding one additional (X,Y) pair $(100,1000000)$. The straight-line fit to the data $(0,3),(0,3),(1,3),(100,1000000)$ has $R^2 \approx 1$. The lesson to learn from this example is that R^2 is only a rough measure of fit. The presence of outliers can make the model appear to be better or worse than it is in actuality. As discussed previously, the coefficient of determination is completely inappropriate if the underlying relationship is highly nonlinear.

10.5 VERIFYING THE MAIN ASSUMPTIONS IN LINEAR REGRESSION

As was discussed at the beginning of the chapter, there are a number of underlying assumptions that must be made before linear regression techniques can be employed. It is imperative that you understand underlying assumptions because you may have to address this issue in order to validate your research. Specifically speaking, if you present results at a conference, you will in all likelihood find that some members of your audience will ask whether you checked your assumptions. This section will help you make these comparisons.

One of the assumptions was that the variance was constant (e.g., regardless of the value of x, the variance is constant). The most common method for checking this assumption is to plot the residuals (e.g., the residual difference between the predicted value of y and the observed value of y, for a given x) vs. the predicted value, or the x variable. These plots are used to check that the magnitude of the residuals does not vary in a systematic way with the predicted variable or the x variable. For example, if the residuals' spread appears to increase (or decrease) as x increases, then the assumption of constant variance is probably not appropriate. When a nonconstant variance is suspected, the data are typically handled by either performing weighted least squares or by transformations, such as Box-Cox transformations. An excellent reference for handling nonconstant variance is Carroll and Rupert (1988). Note that if the assumption is violated, this is usually not considered a fatal error. This is because the estimated slope and intercept are still unbiased, but unnecessarily inaccurate. The resulting population variances would be bigger than necessary. Moreover, the estimated standard errors would be biased and thus inaccurate.

Another assumption is that the straight-line model is an adequate representation of reality. Again, residual plots should be checked. There should be no discernable pattern in the residuals, as you should only see random noise. When there are suspicions that the linear fit is not adequate, then smoothing methods from the appendix or higher-order polynomials may be appropriate. Polynomial models will be discussed later in this chapter.

Finally, the assumption of independent errors is critical for obtaining valid confidence intervals and prediction intervals. Time series techniques can be used to check for serial correlation in the data. See the JMP manual for these tests.

Mosteller and Tukey (1977) give excellent diagnostic procedures and alternative approaches when assumptions appear to be violated.

10.6 COMPARING TWO REGRESSION LINES AT A POINT AND COMPARING TWO REGRESSION PARAMETERS

The reader may be confronted with the question of whether two different regression models are, in fact, statistically dissimilar. Fortunately, using the formulas that we have studied, it is relatively easy to compare (1) two regression lines at the same point x, and (2) regression parameters from two independent experiments. The logic behind the approach is similar to the logic used when we compared two means and, later, two proportions. The general form of the confidence interval and corresponding test will be similar regardless if we are comparing two independent estimates of the slope, or if we are comparing two predictions. If we want to test whether two slopes, b_1 and b_2, are equal, we use the confidence interval

$$\hat{b}_1 - \hat{b}_2 \pm z_{1-\alpha/2} \sqrt{\hat{\sigma}_{b_1}^2 + \hat{\sigma}_{b_2}^2}. \qquad (10.18)$$

If zero is not in this interval, we conclude that the slopes are not equal.

For comparing two independently estimated regression lines at a point x_0 from two different models at a common value, we construct the confidence interval

$$(\hat{a}_1 - \hat{a}_2) + (\hat{b}_1 - \hat{b}_2)X_0 \pm z_{1-\alpha/2}\sqrt{\hat{\sigma}^2_{\hat{a}_1+\hat{b}_1X_0} + \hat{\sigma}^2_{\hat{a}_2+\hat{b}_2X_0}}. \qquad (10.19)$$

If zero is not in that interval, the predictions are said to be different.

10.7 THE REGRESSION DISCONTINUITY DESIGN (RDD)

In many transportation evaluations, there is an abrupt change in policy with a predictor variable or covariate. The predictor may be time, for example, a new speed limit that goes into effect on a specific date, or it could be the age of a (potential) driver. These social policies can be evaluated by fitting a separate regression line to the two groups. For example, suppose drivers at age X_0 are required to have an eye exam in order to keep their driver's license. We can fit a line to the number of tickets for ages X_0-5 to X_0 and a different line to the data for drivers X_0 to X_0+5. We then test to see whether the difference estimate in Equation 10.19,

$$(\hat{a}_1 - \hat{a}_2) + (\hat{b}_1 - \hat{b}_2)X_0 \pm z_{1-\alpha/2}\sqrt{\hat{\sigma}^2_{\hat{a}_1+\hat{b}_1X_0} + \hat{\sigma}^2_{\hat{a}_2+\hat{b}_2X_0}}.$$

contains zero. If the confidence interval contains only negative values, then we might attribute the improvement to the eye exam. There are many other explanations that need to be considered as possible causes of a significant difference. For instance, safety improvements in vehicles or roads could provide alternating explanations. Readers are referred to Campbell and Stanley (1963), Cook and Campbell (1979), Shadish et al. (2002), Sween (1971), and Trochim (1984). These references provide important considerations and an in-depth discussion of the regression discontinuity design. They also contain commonsense guides underlying social program evaluation. For a history of RDD, see Cook (2008).

10.8 MULTIPLE LINEAR REGRESSION

In this section, we will extend simple linear regression to include more than one predictor. For example, traffic congestion depends upon the numbers of highways, entrances, exits, arterial roads, signals, and many other factors. Thus, engineers wanting to predict congestion need tools that can use multiple predictor variables as input. In addition, engineers will often need to explore many alternative models in order to find the one most relevant to their purpose.

This section describes common tools for predicting a single outcome measure using multiple predictors. This includes polynomial regression where multiple predictors are constructed from one original predictor using a power transformation.

Assume that we observe n observations on $p+1$ variables $(Y_i, X_{1i}, X_{2i}, \ldots, X_{pi})$. They are related by the equation

$$Y_i = \beta_0 + \beta_1 X_1 + \beta_2 X_2 + \ldots + \beta_p X_p + \varepsilon_i. \tag{10.20}$$

As in simple linear regression, Y_i is the dependent variable that is to be modeled as a linear function of the p predictor variables $(x_{1i}, x_{2i}, \ldots, x_{pi})$. Just as in simple linear regression, the terms ε_i represent independent and identically distributed (iid) normal random error and have mean of 0 and variance σ^2. The parameter β_0 is the intercept, and the parameters $\beta_1, \beta_2, \ldots, \beta_p$ are the regression slopes for the predictor variables $(X_{1i}, X_{2i}, \ldots, X_{pi})$, respectively. Under these assumptions, the least-squares estimates estimators $\hat{\beta}_0, \hat{\beta}_1, \ldots, \hat{\beta}_p$ are obtained as the parameter values that minimize

$$\sum_{i=1}^{n} \left(Y_i - \beta_0 - \beta_1 X_1 - \ldots - \beta_p X_p \right)^2. \tag{10.21}$$

The variance estimator $\hat{\sigma}^2$ is calculated by the equation

$$\hat{\sigma}^2 = \frac{\sum_{i=1}^{n} \left(Y_i - \hat{\beta}_0 - \hat{\beta}_1 X_1 - \ldots - \hat{\beta}_p X_p \right)^2}{n-p} = \frac{\sum_{i=1}^{n} \left(Y_i - \hat{Y}_i \right)^2}{n-p}. \tag{10.22}$$

The formulas for the standard errors for $\hat{\beta}_0, \hat{\beta}_1, \ldots, \hat{\beta}_p$ are best given using matrices, and they can be found in Ott and Longnecker (2009). Multiple linear regression also includes polynomial regression. Note that in the model

$$Y = \beta_0 + \beta_1 X + \beta X^2 + \ldots + X^p + \varepsilon \qquad (10.23)$$

is a special case of multiple linear regression. In polynomial regression, $X_1 = X$, $X_2 = X^2$, ..., $X_p = X^p$. The term *linear* in linear regression refers to the linearity of the model in the regression parameters, or betas. Multiple linear regression models are not necessarily linear in the Xs.

To illustrate this concept, consider the hours of traffic delay for the city of Los Angeles for the years 1982 to 2004. The predictor variables are the population, the average daily miles driven on available arterial streets, and the miles of freeway that were available. A scatter plot matrix is shown in Figure 10.6.

We see from the example that the average daily miles driven on principal arterial streets, the average miles driven on freeways, and the population in Los Angeles are linearly related to the annual hours of delay and vice versa. We now use Equation 10.20 to model the annual hours of

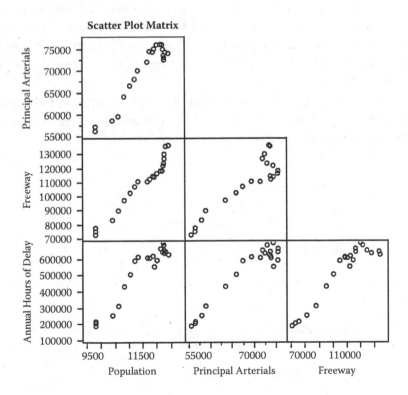

FIGURE 10.6 Scatter plot matrix.

delay as a linear function of the other variables. The estimated parameters, their standard errors, t-tests, and corresponding p values are shown in Figure 10.7.

The parameter estimates for $\beta_0, \beta_1, \beta_2$, and β_3 are given in the first column of Figure 10.7. The next column contains their estimated standard error. The third column contains the corresponding t-test for testing the hypothesis that the parameter is zero vs. the alternative hypothesis that the parameter is not zero. We notice that only one p value is significant at the $\alpha = .05$ level (e.g., average miles driven on principal arterial streets with a t value of 2.91). It is tempting to refit the model using only the intercept and the number of principal arterial streets as the predictors. However, when we look at the matrix of the correlations among the predictors in Figure 10.8, we see that the predictors are highly correlated.

High correlation among predictor variables is known as *collinearity* of predictors. It causes the standard errors of the collinear predictors and the corresponding p values to be inflated. In this situation, we can remove any two of the three x variables, and the remaining slope, in all likelihood, will be statistically significant. More importantly, not removing two of the variables will make the model appear to be more useful than it actually is.

Lastly, it should be pointed out that it is not uncommon to have collinearity in actual engineering data. In fact, it would be rare to have correlation values close to zero for observational studies.

▼ Parameter Estimates

Term	Estimate	Std Error	t Ratio	Prob>|t|
Intercept	−722203.7	437626.1	−1.65	0.1162
Population	−30.56175	85.11012	−0.36	0.7237
Freeway	4.1640404	2.67982	1.55	0.1376
Principal Arterials	16.542325	5.691256	2.91	0.0094*

FIGURE 10.7   Parameter estimates.

Correlations

	Population	Principal Arterials	Freeway
Population	1.0000	0.9629	0.9764
Principal Arterials	0.9629	1.0000	0.9163
Freeway	0.9764	0.9163	1.0000

FIGURE 10.8   Correlation matrix for the predictors.

10.8.1 Confidence Intervals on Regression Parameters

Confidence intervals can be calculated for the $k (= p + 1)$ regression parameters (including the intercept) using the formula

$$\hat{\beta}_i \pm t(1 - \alpha / 2, n - k) t_{n-k, 1-\alpha/2} \qquad (10.24)$$

More advanced textbooks, such Johnson and Wichern (2007), give formulas for confidence ellipsoids. Confidence ellipsoids give joint confidence regions for the p-estimated regression parameters that have a minimum area.

10.8.1.1 Extrapolation

It is extremely dangerous to use predictors outside of the joint range of experimental data. For observational studies where data are typically collinear, this is very important. For example, suppose that two x variables were practically equal everywhere in the data used to fit the model (and hence nearly perfectly collinear). Then any prediction conducted where the same two x variables were not nearly equal would be outside the range of the experimental data, even if they were not extreme values compared to the data used to fit the model. It is dangerous to extrapolate using x data that are not typical of the data used to fit the model.

10.9 VARIABLE SELECTION FOR REGRESSION MODELS

Given that there are multiple variables that can be used in any model (some of which may be correlated), a natural question is: How do you identify a reasonable model? A brute force approach would be to calculate each possible model, remove the ones that are clearly incorrect (because they defy common sense, for example), and then use some predetermined criteria to choose the best one (e.g., highest R^2). Obviously, this would be time consuming. Fortunately, the statistical packages have automatic techniques that can be used, and these applications are the focus of this section.

Stepwise modeling procedures usually give three choices: step up, step down, and mixed procedures. In JMP, the step-up method starts adding predictor variables one at a time until a criterion of choice reaches a level of choice. The step-down method starts with all predictor variables in the fitted model and removes them one at a time until a criterion of choice reaches a level of choice. Finally, a mixed method allows adding and subtracting predictor variables until a criterion of choice reaches a level of choice. In addition when the number of predictors is small, JMP allows searching all possible subsets of predictors.

When the sample size is much greater than the number of predictors, the authors think it is best to start with a step-down approach. When the sample size is small or about the same size as the number of predictor variables, then the authors' preference is to start with a step-up approach. For this example, all three approaches arrive at the same optimal model, but this does not always happen. When the stepwise model selection methods choose different models, then care must be taken in choosing the best model. Chosen models by different selection methods typically differ when there is multicollinearity. It should be noted that each of the models will be statistically significant based on the criteria defined *a priori* by the modeler. Thereby, if the modeler requested a 95% of level of significance, then all the identified models will have variables that are statistically significant at this level. As previously stated, there is no set way to choose the best model, and hence this decision is based on the engineer's judgment. One useful principle, known as Occam's razor, states that when comparing two models that work equally well, the simplest model should be chosen. It is very important that users of a developed model should be informed about the existence of alternative models that work nearly as well as the chosen model. While alternative models may fit nearly equally well on the data set used to test the models, they may not fit equally well on more extensive data.

It cannot be stressed enough that there is no way to identify the best model, and as such, the engineer is often responsible for making this decision. For example, a model with four input parameters may be marginally better than a model with two input parameters based on statistical measurements. However, if it costs a significant amount of money to obtain observations on extra parameters, the simpler model might be preferred. By the same token, it may be that one of the extra parameters is a policy variable (e.g., price of fuel) that is needed for a particular analysis. In this case, it may be more reasonable to select the more complex model.

The package JMP has stepwise (forward, backward, and mixed) model selection methods available in its "Fit Model" procedure under the "Analysis" menu. The JMP stepwise procedures give several methods for judging the quality of the fit. These include R^2, C_p, adjusted R^2, and AIC collected. AIC corrected is a statistical method that balances the choice of model by adding a well chosen penalty for the number of predictors used to the residual sum of squares (see Figure 10.9). In addition,

Current Estimates							
SSE	DFE	MSE	RSquare	RSquare Adj		Cp	AICc
3.824e+10	19	2.0126e+9	0.9383	0.9319		2.1289419	540.8602

Lock	Entered	Parameter	Estimate	nDF	SS	"F Ratio"	"Prob>F"
✓	✓	Intercept	-872777.07	1	0	0.000	1
☐	☐	Population	0	1	2.72e+8	0.129	0.72371
☐	✓	Freeway	3.32673414	1	1.34e+10	6.650	0.0184
☐	✓	Principal Arterials	14.9307555	1	3.84e+10	19.075	0.00033

FIGURE 10.9   Estimated parameters from a stepwise fit in JMP.

p values are provided. The interpretation of R^2 is similar to that of the simple linear regression R^2. The C_p statistic is a frequently used technique, and the usual recommendation is to choose the model that has C_p equal to the number of predictor variables plus 1 for the intercept. The adjusted R^2 is an R^2 value with a correction for degrees of freedom. The uncorrected R^2 increases as more predictors are added to the model. However, the adjusted R^2 value does not necessarily increase as parameters are added to the model. Finally, a very popular measure of fit is the AIC criterion. The model with the lowest AIC value is chosen as the best. For additional details, see the JMP manual, and Sheather (2009).

For the Los Angeles data, the model using JMP with the lowest AIC has annual hours of delay predicted by the model with predictors—the average daily miles driven on principal arterial streets and average daily miles driven on freeways—as the predictor variables. These results are shown in Figure 10.9. The parameters forced into the model have a checkmark in the "lock" column, and those chosen by the algorithm have a checkmark in the entered column. While the p values are given for each of the parameters, they are not statistically valid. This is because they do not account for the multiple numbers of tests performed in choosing a model. Readers can find more details about the stepwise regression method in a number of excellent texts, such as those written by Neter et al. (1996), Miller (1990), and Sheather (2009).

10.10 ADDITIONAL COLLINEARITY ISSUES

When predictor variables are linearly related, as in the Los Angeles example, the parameter estimates will have an inflated standard error and correspondingly wide confidence intervals. It is important for transportation engineers to understand that modeling when predictor variables are collinear is a tenuous process. Small changes in the values for a predictor can lead the stepwise regression procedure to choose a different model.

When data sets are large, it is usually wise to randomly choose subsets of the data to model. For example, a modeler could choose ten random subsets, each using 90% of the original data. If any of the resulting ten models are different in important ways, then the modeler needs both to investigate the differences and to explain these discrepancies to users of the model obtained from the full data. However, a full discussion of this approach is beyond the scope of the book. The authors recommend that the reader seek professional statistical assistance in these situations.

10.11 CONCLUDING REMARKS

The focus of this chapter was on modeling the linear relationships between a dependent variable (y) and one or more predictor variables (X_i). These types of models are often used in transportation and are found in many applications. Given that the various assumptions—constant variance, iid data, and normality—are valid, these models can be very useful. The ways of developing confidence intervals and prediction intervals for the predicted values were also covered in this chapter. The former are related to how well we can predict the mean response, while the latter pertain to our ability to predict a particular observation. For example, we may want to predict speed for a given speed limit. We would use the CI to measure how well we can predict the average or mean speed and the PI for a range of observations we might expect.

Model building was also a major focus of this chapter. We explained that there are a number of criteria that can be used to identify whether a model is useful. In addition, methods for finding the best models automatically were illustrated. However, at the end of the day, model building is as much an art as a science, as it is up to the modeler to use her judgment to identify the best model. The user must be prepared to justify her selection based on statistical theory. Additionally, she should be prepared to test the underlying assumptions to demonstrate that linear regression is appropriate for the given situation.

In summary, linear regression (both simple and multiple) is a very powerful tool and is often used in transportation applications. However, the reader is cautioned to understand the underlying assumptions behind the model so that the technique can be used correctly. This chapter's homework problems will illustrate some of the common mistakes.

HOMEWORK PROBLEMS

1. Double-loop data for San Antonio is contained in the data file corresponding to Figure 3.1.

   a. Plot occupancy (Y) vs. speed (X).

   b. Fit a line to the data.

   c. Is the slope of the line significantly different from zero? With what confidence level?

   d. What is a good estimate of the error standard deviation? How many degrees of freedom does it have?

   e. What is R^2?

   f. Plot the residuals. Do any assumptions appear to be violated?

   g. Split the data into those observations with speeds less than 30 mph, and those with speeds equal to or greater than 30 mph. Fit both sets separately and compare the fits. What can you conclude?

2. A field gauge used to measure asphalt density is compared to a lab instrument. The data are in the file called "field gauge." Engineers want to know if the field gauge is equivalent to the lab instrument.

   a. If instruments are equivalent, then a zero reading on one instrument should be zero on the other, at least on average. Is there evidence to support that this is true for the lab and field gauge?

   b. If instruments are equivalent, then the scale of measurement should be the same for both instruments. That is, if the instruments are equivalent, then the straight-line fit should give an intercept of zero and a slope equal to 1 up to uncertainty levels. Is that true for this example?

   c. Give a confidence interval for ρ, the population correlation coefficient.

   d. Mobility data for Akron, Ohio, are contained in the data file called "problem 10.3." For the six mobility variables in the file, compute their correlations. Which ones are significantly different from zero?

    e. Using stepwise regression, build a model to predict public transportation usage.

    f. For your model, what is R^2? Is your model reasonable?

REFERENCES

Campbell, D. T., and J. C. Stanley. 1963. *Experimental and quasi-experimental designs for research*. Boston: Houghton Mifflin.

Carroll, R. J., and D. Rupert. 1988. *Transformation and weighting in regression*. New York: Chapman & Hall.

Cook, T. D. 2008. "Waiting for life to arrive": A history of the regression-discontinuity design in psychology, statistics and economics. *Journal of Econometrics*, 142, 636–54.

Cook, T. D., and D. T. Campbell. 1979. *Quasi-experimentation: Design and analysis for field settings*. Chicago: Rand McNally.

Johnson, R. A., and D. W. Wichern. 2007. *Applied multivariate statistical analysis*. 7th ed. Englewood Cliffs, NJ: Prentice Hall.

Miller, A. J. 1990. *Subset selection in regression*. London: Chapman & Hall.

Mosteller, F., and J. W. Tukey. 1977. *Data analysis and regression: A second course in statistics*. Addison-Wesley Series in Behavioral Science. Reading, MA: Addison-Wesley.

Neter, J., M. H. Kutner, C. J. Nachtsheim, and W. Wasserman. 1996. *Applied linear statistical models*. 4th ed. Burr Ridge, IL: Irwin.

Ott, R., and M. Longnecker. 2009. *An introduction to statistical methods and data analysis*. 6th ed. Belmont, CA: Cengage/Brooks/Cole.

Schrank, D. T., and T. Lomax. 2007. *The urban mobility report*. Texas Transportation Institute, College Station, TX.

Shadish, W. R., T. D. Cook, and D. T. Campbell. (2002). *Experimental and quasiexperimental designs for generalized casual inference*. Boston: Houghton Mifflin Company.

Sheather, S. J. 2009. *A modern approach to regression with R*. New York, New York: Springer.

Sween, J. A. 1971. *The experimental regression design: An inquiry into the feasibility of nonrandom treatment allocation*. Evanston, IL: Northwestern University.

Trochim, W. M. K. 1984. *Research design for program evaluation*. Beverly Hills, CA: Sage.

CHAPTER 11

Regression Models
for Count Data

11.1 INTRODUCTION

In Chapter 10 we looked at linear regression models that are commonly used in transportation to identify the effects of independent variables (whether continuous or discrete) on an important dependent variable (that is, continuous). For example, we might be interested in the speed on a roadway as a function of the traffic volume, the number of lanes, the roadway density, the type of driver, etc. However, sometimes in transportation the dependent variable is discrete in nature. One common example would be the crash frequency counts in safety analysis. The modeling of safety is important to engineers and decision makers.

In safety analysis, we are often interested in finding the relationship between crashes and roadway characteristic variables such as number of lanes, lane width, shoulder width, median type, median width, etc., and traffic volumes. Crashes are often examined in terms of crash counts, and because of the nature of crash counts, they are discrete and rare. They usually do not follow a normal distribution and, in general, it has been found that the variance increases as the crash count increases. Consequently, it is usually inappropriate to model untransformed crash count as a linear regression model with a normal error distribution because some of the underlying model assumptions such as normality and constant variance are seriously violated. Note that, as discussed in Chapter 10, it is possible that the crash counts can be transformed so that the distribution of the transformed variable becomes closer to a normal distribution with constant variance. In this situation, a linear regression model may be employed to develop a prediction equation. However, a transformation does not always provide a satisfactory result; namely, transformed crash counts may still not be close enough to a normal distribution.

Alternative approaches, which are the focus of this chapter, involve the use of Poisson regression or a negative binomial regression models because they are specifically designed to model discrete events. These models can describe the relationship between crash counts and roadway characteristic variables without the need to first transform the data. These models belong to a category of the generalized linear models (see, e.g., McCullagh and Nelder, 1989), and they are widely used nowadays to model crash count data.

11.2 POISSON REGRESSION MODEL

A Poisson regression model can be used to describe the number of crashes, Y, occurring for a fixed time interval (such as a year or a month) on several roadway segments or on a roadway segment over time intervals. In Poisson regression, the dependent variable, Y, is assumed to follow a Poisson distribution with the mean μ, which is dependent on explanatory variables, such as the roadway geometric characteristic variables or traffic volumes. Recall that a Poisson distribution can be expressed as

$$\Pr(Y=y)=\frac{\exp(-\mu)\mu^y}{y!}, \quad y=0,1,2,\cdots.$$

In Poisson regression, it is assumed that the log of mean μ is a linear function of a set of covariates, that is,

$$\log(E(Y))=\log(\mu)=\beta_0+\beta_1 x_1+\cdots+\beta_p x_p,$$

or equivalently,

$$\mu=\exp(\beta_0+\beta_1 x_1+\cdots+\beta_p x_p).$$

The covariates are often some combination of control variables, such as roadway width, shoulder width, etc., and other variables such as traffic volume.

Let's denote the crash count per year at the ith roadway segment by Y_i $(i=1,\cdots,n)$. Then the Poisson regression model for Y_i can be written as

$$Y_i \sim \text{Poisson}(\mu_i)$$

where

$$\mu_i = \exp\left(\beta_0 + \sum_j \beta_j x_{ij}\right)$$

and Y_is are assumed to be independent given μ_i. The exponential coefficient for x_j, $\exp(\beta_j)$, $j=1,\cdots,p$, indicates the effect of unit change of x_j on crashes. The percent change in crashes can also be measured by $\{\exp(\beta_j)-1\}\times 100\%$.

The regression parameters, $\beta_0,\beta_1,\cdots,\beta_p$, are often estimated by the maximum likelihood estimation method (which maximizes the likelihood or the log likelihood with respect to parameters (here βs)). A brief description of the maximum likelihood estimation method is provided in the appendix to this chapter.

The log likelihood (LL) for the Poisson regression is given as

$$LL = \sum_{i=1}^{n} \{y_i \mathbf{x}_i \beta - \exp(\mathbf{x}_i \beta) - \log(y_i!)\}$$

where $\mathbf{x}_i \beta = \beta_0 + \beta_1 x_{i1} + \cdots + \beta_p x_{ip}$. Maximum likelihood estimation can be implemented by a number of statistical software packages. For example, the GENMOD and GENLOG procedures can be used to fit Poisson regression models in SAS and SPSS, respectively. In JMP, the Poisson regression model can be fitted in the "Model Specification" or "Fit Model," window as shown in Figure 11.1. This window is in turn a subset of the "Analyze" menu. It can be seen in the upper right corner of the menu that "Generalized Linear Model," "Poisson," and "Log" are selected for the "Personality," "Distribution," and "Link Function," respectively.

One popular application is to use a Poisson regression model for crash rate data where the crash rate is a count of crashes divided by exposure. A common measure of exposure is in million vehicle-miles of travel per year. An example calculation of exposure for a given link (or segment) i is shown below:

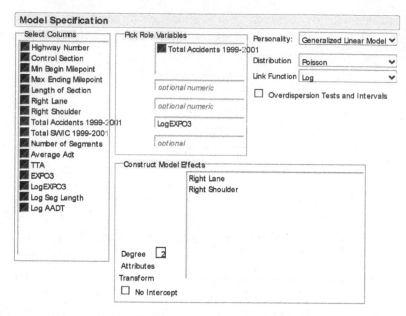

FIGURE 11.1   JMP fit model window for model specification.

$$\text{exposure}_i = \frac{ADT_i \times 365 \times \text{Length}_i}{10^6}.$$

In Poisson regression, the rate data can be handled by using the original crash counts (the numerator of the crash rate) as a dependent variable while having the log of exposure as an offset variable, as follows:

$$\log(E(Y_i)) = \log(\text{exposure}_i) + \beta_0 + \beta_1 x_{i1} + \cdots + \beta_p x_{ip}.$$

An offset variable is a regression variable with a constant coefficient of 1 for each observation.

11.3 OVERDISPERSION

Recall that the Poisson distribution is completely characterized by one parameter, the mean (μ), which is also equal to the variance. However, there are many cases where the observed variance of crash counts is greater than the mean. For example, this may occur because of heterogeneity of roadway segments, a failure to include important explanatory variables in the model, and for myriad other reasons. As an aside, crashes are related to (1) roadway characteristics, (2) vehicle characteristics, and (3) driver characteristics. Obtaining accurate information on the latter is practically impossible, and consequently, it is fairly common to not include all the relevant explanatory variables in crash models.

The above phenomenon, where the variance is greater than the mean, is called overdispersion. In this situation a Poisson model would not be appropriate and the modeler must account for the excess variation. If overdispersion in the data is ignored when it is actually present, then the standard errors of the estimated regression coefficients under Poisson regression models are underestimated and erroneous inferences could be made (e.g., the variables might be incorrectly declared to be significant when they are, in fact, not). A common solution to this problem is to use negative binomial regression models, which will be introduced in the next section.

As an aside, when overdispersion is not very severe, an overdispersed Poisson model can also be used, which introduces a scale parameter to correct the estimated covariance matrix under the Poisson model. It does this by inflating the standard errors by the value of the scale parameter (see, e.g., Cohen, 2003). This technique will be illustrated later in the chapter.

11.4 ASSESSING GOODNESS OF FIT OF POISSON REGRESSION MODELS

One of the popular measures used to assess the fit of Poisson regression models is the deviance, which is a measure of unexplained variation similar to the residual sum of squares in multiple linear regression. It is defined as

$$\text{Deviance}(\hat{\mu}, Y) = 2\{\log L(Y, Y) - \log L(\hat{\mu}, Y)\}$$

where $\log L(\hat{\mu}, Y)$ is the log likelihood under the fitted model $\hat{\mu}$ and $\log L(Y, Y)$ is the log likelihood for the largest model having each observation y as a unique estimate of μ.

For the Poisson model the deviance can be expressed as follows:

$$\text{Deviance}(\hat{\mu}, Y) = 2 \sum_{i=1}^{n} \{Y_i \log(Y_i / \hat{\mu}_i) - (Y_i - \hat{\mu}_i)\}.$$

If our model fits the data well, it is expected that the deviance has an approximate chi-square distribution with degrees of freedom (DF) $n - q$, where n is the number of observations and q is the number of parameters (i.e., $q = p + 1$, including an intercept) in the fitted model $\hat{\mu}$. Thus, *deviance*/DF should be close to 1 if the model fits the data well. A value of *deviance*/DF much greater than 1 suggests that the model is misspecified or there is an overdispersion problem in the data. A value of *deviance*/DF much less than 1 also indicates misspecification of the model or an underdispersion problem (the phenomenon that the variance is less than the mean, which also violates an underlying assumption for the Poisson distribution) in the data (Sheather, 2009). Caution needs to be exercised with using the deviance as a goodness-of-fit measure when the data set being analyzed is sparse (when there is a large number of observed zeros in the data) because comparison of the deviance with a chi-square distribution may be invalid in such cases, as discussed in Boyle et al. (1997). A simulation approach was suggested by Boyle et al. (1997) as an alternative method of assessing model fit for the sparse data.

Another popular measure of fit of the model is the Pearson chi-square statistic, which is defined as the squared difference between the observed and predicted values divided by the variance of the predicted value summed over all observations in the model:

$$X^2 = \sum_{i=1}^{n} \frac{\left(Y_i - \hat{\mu}_i\right)^2}{Var\left(\hat{\mu}_i\right)}.$$

Similar to the deviance, the Pearson chi-square has an approximate chi-square distribution with degrees of freedom $n - q$ if the model fits the data well. Thus, a value of Pearson chi-square/DF close to 1 indicates a good model fit, while a value much greater than 1 or less than 1 may suggest that there is an overdispersion or underdispersion problem in the data.

Example 11.1

Researchers are interested in finding the relationship between crashes on rural two-lane highways and the width of the right lane and the width of the right shoulder. A subset of Texas on-system crashes for the years 1999–2001 was obtained for the analysis (Fitzpatrick et al., 2005). There are 2,729 two-lane roadway segments in the database, and crashes are examined in terms of the variable "total crashes for three years." Total crashes are the total number of all crashes (includes all types of crashes) during the three-year observation period on each roadway segment. In this example, the dependent variable, Y, is "total crashes for three years," and the independent variables of interest are right lane (width in feet) and right shoulder (width in feet). The exposure variable is also introduced into the model to account for differences in segment length and roadway volume across the different roadway segments. The exposure variable, EXPO3, is defined as

EXPO3 = exposure in million vehicle-miles of travel for 3 years

$$= (AADT)(365)(3)(\text{segment length})(10^{-6}).$$

A Poisson regression model of the following functional form is fitted to the data:

E(total crashes for 3 years) = EXPO3 $\exp(\beta_0 + \beta_1$ right lane $+\beta_2$ right shoulder)

The GENMOD procedure in SAS is used to fit Poisson regression models using log(EXPO3) as an offset variable. Table 11.1 contains the SAS output obtained by using PROC GENMOD as follows:

```
PROC GENMOD DATA=RSBS_2L_NM_LS;
MODEL Total_Crashes_for_3_years = Right_Lane Right_Shoulder
/ DIST=Poisson LINK=Log OFFSET=LogEXPO3;
RUN;
```

It can be observed from Table 11.1 that both right lane width and right shoulder width have statistically significant influence. The negative sign for the coefficients for them suggests that highway segments that have a wider right lane or wider right shoulder experience lower crash rates, all else being equal. This result would be consistent with expectation. The exponential coefficient for right lane, $\exp(-0.1104) = 0.8955$, indicates how much the crash rate changes when the right lane width increases by a foot, i.e., the crash rate is reduced by a factor of 0.8955. The percent change in crash rate associated with a foot increase in the width of the right lane is $\{\exp(-0.1104)-1\}\times100\% = -10.45\%$, i.e., 10.45% less. A similar interpretation can also be made for the coefficient of right shoulder width.

The prediction equation obtained by using the estimated coefficients in Table 11.1 can be given as follows:

E(total crashes for 3 years) = EXPO3 exp(1.4018 − 0.1104 right lane − 0.0664 right shoulder).

Note that the goodness of fit information shows that the deviance and Pearson chi-square divided by degrees of freedom somewhat exceed 1.0. Remember that ratios (value/DF) close to 1.0 indicate that the model is adequate and overdispersion is not likely to be a problem. In this case, while the ratios do not exceed 1.0 seriously, they might still cast some doubt about validity of the Poisson model. In other words, the ratios suggest that there is a greater variability among total crash counts than would be expected for Poisson distribution. In technical terms, there may be an issue of overdispersion in the model.

A similar output can also be obtained by running the generalized linear model procedure in JMP given in Figure 11.1. The resulting JMP output is given in Table 11.2.

TABLE 11.1 Model Information for Total Crash Data with a Poisson Regression Implemented by SAS PROC GENMOD

```
                     The GENMOD Procedure
                      Model Information

Data Set             WORK.RSBS_2L_NM_LS
Distribution         Poisson
Link Function        Log
Dependent Variable   Total_Crashes_for_3_years Total Crashes for 3 years
Offset Variable      LogEXPO3 LogEXPO3

        Number of Observations Read    2729
        Number of Observations Used    2729

        Criteria For Assessing Goodness Of Fit

Criterion              DF       Value      Value/DF
Deviance             2726    4156.9730      1.5249
Scaled Deviance      2726    4156.9730      1.5249
Pearson Chi-Square   2726    5666.3015      2.0786
Scaled Pearson X2    2726    5666.3015      2.0786
Log Likelihood                127.9287
```

(Continued)

Algorithm converged.

Analysis Of Parameter Estimates

Parameter	DF	Standard Estimate	Error	Wald 95% Confidence Limits		Chi-Square	Pr > ChiSq	
Intercept	1	1.4018	0.2493	0.9131	1.8905	31.61	<.0001
Right_Lane	1	-0.1104	0.0229	-0.1553	-0.0655	23.26	<.0001
Right_Shoulder	1	-0.0664	0.0045	-0.0753	-0.0575	214.67	<.0001
Scale	0	1.0000	0.0000	1.0000	1.0000		

NOTE: The scale parameter was held fixed.

TABLE 11.2　Model Information for Total Crash Data with a Poisson Regression Implemented by JMP Generalized Linear Model Procedure

Generalized Linear Model Fit

Offset: LogEXPO3
Response: Total crashes for 3 years
Distribution: Poisson
Link: Log
Observations (or Sum Wgts) = 2,729

Whole Model Test

Model	−Log Likelihood	L-R Chi-Square	DF	Probability > Chi-Square
Difference	234.646834	469.2937	2	<.0001
Full	3885.95419			
Reduced	4120.60102			

Goodness of Fit Statistic

Goodness of Fit Statistic	Chi-Square	DF	Probability > Chi-Square
Pearson	5666.302	2,726	<.0001
Deviance	4156.973	2,726	<.0001

Parameter Estimates

Term	Estimate	Standard Error	L-R Chi-Square	Probability > Chi-Square	Lower CL	Upper CL
Intercept	1.4017893	0.2493393	32.840435	<.0001	0.9154081	1.8921311
Right lane	−0.110404	0.0228931	24.157108	<.0001	−0.155487	−0.06581
Right shoulder	−0.066405	0.0045322	212.9771	<.0001	−0.075288	−0.057522

Example 11.2

Because overdispersion in the data of Example 11.1 seems possible, an over-dispersed Poisson model that introduces a scale parameter to correct the estimated covariance matrix under the Poisson model may be tried. The scale parameter can be estimated by the square root of *deviance*/DF (by using SAS PROC GENMOD with an option "scale = *d*") or the square root of Pearson chi-square/DF (by using SAS PROC GENMOD with an option "scale = *p*"). The result from the overdispersed Poisson regression with the scale parameter estimated by the square root of *deviance*/DF is presented in Table 11.3, and was obtained by using SAS PROC GENMOD as follows:

```
PROC GENMOD DATA=RSBS_2L_NM_LS;
MODEL Total_Crashes_for_3_years = Right_Lane Right_Shoulder
/ DIST=Poisson scale=d LINK=Log OFFSET=LogEXPO3;
RUN;
```

It can be seen from the output in Table 11.3 that the scaled deviance is now held fixed to 1 and the scale parameter is estimated as 1.2349. Also note that the parameter estimates for intercept and right lane and right shoulder have not changed. However, their standard errors are now inflated by the value of the scale parameter. Although the effects of right lane and right shoulder remain statistically significant, it is important to note that there is no guarantee that this will always happen. In other words, the scale parameter will not affect the point estimates of the regression coefficients. However, the correspond-ing uncertainty estimates (standard errors) will get larger, which reflects the fact that there is more variability in the data than there would be in a standard Poisson model. The CI and PI intervals (if calculated) would be greater than those of the standard Poisson model.

The overdispersed Poisson regression can also be run in JMP by selecting the option "Overdispersion Tests and Intervals" from the "Fit Model" window given in Figure 11.1. In JMP, however, the scale parameter is always estimated by the square root of Pearson chi-square/DF (not by *deviance*/DF). The result-ing standard error estimates from JMP would be equal to those of SAS if an option "scale = *p*" were used in running SAS PROC GENMOD. However, they might be different if an option "scale = *d*" were used in SAS. Table 11.4 con-tains the JMP output of fitting an overdispersed Poisson regression model to the same data.

11.5 NEGATIVE BINOMIAL REGRESSION MODEL

Recall that the critical assumption in the Poisson distribution is that the mean (μ) is equal to the variance. However, it is well documented in the literature (see, e.g., Lord et al., 2005) that this is often violated for crash data, which means that overdispersion is an issue in model development. To account for overdispersion, the negative binomial (NB) model is often used. The NB model introduces an additional parameter, k, called the

TABLE 11.3 Model Information for Total Crash Data with an Overdispersed Poisson Regression Implemented by SAS PROC GENMOD

```
                      The GENMOD Procedure
                        Model Information

Data Set             WORK.RSBS_2L_NM_LS
Distribution            Poisson
Link Function           Log
Dependent Variable   Total_Crashes_for_3_years Total Crashes for 3 years
Offset Variable         LogEXPO3 LogEXPO3

     Number of Observations Read    2729
     Number of Observations Used    2729

       Criteria For Assessing Goodness Of Fit

Criterion              DF      Value       Value/DF
Deviance              2726   4156.9730      1.5249
Scaled Deviance       2726   2726.0000      1.0000
Pearson Chi-Square    2726   5666.3015      2.0786
Scaled Pearson X2     2726   3715.7658      1.3631
Log Likelihood                83.8913
```

Algorithm converged.

Analysis Of Parameter Estimates

Parameter	DF	Estimate	Standard Error	Wald 95% Confidence Limits		Chi-Square	Pr > ChiSq	
Intercept	1	1.4018	0.3079	0.7983		2.0053	20.73	<.0001
Right_Lane	1	-0.1104	0.0283	-0.1658		-0.0550	15.25	<.0001
Right_Shoulder	1	-0.0664	0.0056	-0.0774		-0.0554	140.77	<.0001
Scale	0	1.2349	0.0000	1.2349		1.2349		

NOTE: The scale parameter was estimated by the square root of deviance/DOF.

TABLE 11.4 Model Information for Total Crash Data with an Overdispersed Poisson Regression Implemented by JMP Generalized Linear Model Procedure

Generalized Linear Model Fit

Offset: LogEXPO3
Overdispersion parameter estimated by Pearson Chisq/DF
Response: Total crashes for 3 years
Distribution: Poisson
Link: Log
Observations (or Sum Wgts) = 2,729

Whole Model Test

Model	–Log Likelihood	L-R Chi-Square	DF	Probability > Chi-Square
Difference	112.886204	225.7724	2	<.0001
Full	1869.49301			
Reduced	1982.37921			

Goodness of Fit

Statistic	Chi-Square	DF	Probability > Chi-Square	Overdispersion
Pearson	5666.302	2726	<.0001	2.0786
Deviance	4156.973	2726	<.0001	

Parameter Estimates

Term	Estimate	Standard Error	L-R Chi-Square	Probability > Chi-Square	Lower CL	Upper CL
Intercept	1.4017893			.	0.7023959	2.1095287
Right lane	-0.110404	0.0330059	11.621739	0.0007	-0.175513	-0.046321
Right shoulder	-0.066405	0.0065343	102.46111	<.0001	-0.0792	-0.053587

negative binomial dispersion parameter, to account for excess variation. In negative binominal regression, the dependent variable, Y, is assumed to follow a negative binominal distribution with the mean μ, which is dependent on explanatory variables such as roadway geometric characteristic variables or traffic volumes. The negative binomial can be derived as a gamma mixture of Poisson by assuming that the mean of Poisson distribution varies across the population of individuals according to a gamma distribution. For the negative binomial regression model, we can write:

$$Y_i | \mu_i \sim \text{Poisson}(\mu_i)$$

where

$$\mu_i = \exp(\beta_0 + \beta_1 x_{i1} + \cdots \beta_p x_{ip} + b_i) = \exp(\beta_0 + \beta_1 x_{i1} + \cdots \beta_p x_{ip}) \exp(b_i)$$

and b_i is an error term corresponding to an outcome-specific latent effect. Assuming that $\exp(b_i)$ is gamma distributed with a mean equal to 1 and variance equal to k, a closed-form expression for the marginal distribution of Y_i can be obtained as the negative binomial:

$$\Pr(Y_i = y_i) = \frac{\Gamma(k^{-1} + y_i)}{\Gamma(y_i + 1)\Gamma(k^{-1})} \left(\frac{k^{-1}}{k^{-1} + \mu}\right)^{1/k} \left(\frac{\mu}{k^{-1} + \mu}\right)^{y_i}, \quad y_i = 0, 1, 2, \cdots$$

where Γ is the gamma function. Note that the functional form for the mean of Y_i remains the same as that for the Poisson regression model. Thus, the expected number of crashes can still be represented as

$$\mu_i = \exp(\beta_0 + \beta_1 x_{i1} + \cdots + \beta_p x_{ip}).$$

However, the variance of Y_i is $\mu_i + k\mu_i^2$, which is larger than that of the Poisson regression model for a positive value of k. It needs to be noted that the negative binomial dispersion parameter, k, is not the same as the dispersion parameter in generalized linear models denoted by ϕ in McCullagh and Nelder (1989), and must be estimated. The SAS PROC GENMOD procedure provides the maximum likelihood estimate for k in addition to the estimates for the regression parameters. Note that the

Poisson regression model corresponds to the negative binominal regression model when the value of k is 0.

Example 11.3

Recall that the crash data in Example 11.1 revealed an overdispersion problem. Although an overdispersed Poisson model can be used to correct the estimated covariance matrix when overdispersion is modest, an negative binomial regression model is, in general, a more preferred model when the data are overdispersed. Table 11.5 contains the SAS output for the negative binominal regression obtained by using SAS PROC GENMOD as follows:

```
PROC GENMOD DATA=RSBS_2L_NM_LS;
MODEL Total_Crashes_for_3_years = Right_Lane Right_Shoulder
/ DIST=NB LINK=Log OFFSET=LogEXPO3;
RUN;
```

From Table 11.5, it can be seen that the effects of right lane and right shoulder are statistically significant, and the signs of the estimated coefficients are consistent with expectation. Note that the standard errors for the coefficient estimates are larger than those of the Poisson regression model, which is a consequence of accounting for overdispersion in the data. The negative binomial dispersion parameter was estimated to be 0.4566. The estimated equation for the expected total crashes is as follows:

E(total crashes for 3 years) = EXPO3 exp(1.4018 − 0.0798 right lane − 0.0663 right shoulder).

From the "criteria for assessing goodness-of-fit" (Table 11.5), it can be seen that the negative binomial model fits the data very well because it has a deviance of 2,571 as opposed to 4,157 for the Poisson regression model, with 2,726 degrees of freedom.

Note that in Examples 11.1 to 11.3, the traffic volume variable (average annual daily traffic (AADT)) was used simply for an offset variable (as part of the exposure variable) rather than as a predictor variable of the regression models. The underlying assumption for using AADT as an offset variable is that the relationship between crashes and AADT is linear. That is, crashes increase with AADT in a linear manner on rural roads in Texas (for the AADT volumes in the study). Intuitively, the relationship between crashes and AADT can be nonlinear. Therefore, it would be better to use AADT as a predictor variable rather than as an offset variable. Because we model log(mean crashes) in Poisson or negative binomial regressions, log(AADT) is often used as a predictor variable.

TABLE 11.5   Model Information for Total Crash Data with Negative Binominal Regression Implemented by SAS PROC GENMOD

The GENMOD Procedure

Model Information

Data Set	WORK.RSBS_2L_NM_LS
Distribution	Negative Binomial
Link Function	Log
Dependent Variable	Total_Crashes_for_3_years Total Crashes for 3 years
Offset Variable	LogEXPO3 LogEXPO3

Number of Observations Read 2729
Number of Observations Used 2729

Criteria For Assessing Goodness Of Fit

Criterion	DF	Value	Value/DF
Deviance	2726	2571.3502	0.9433
Scaled Deviance	2726	2571.3502	0.9433
Pearson Chi-Square	2726	3660.3908	1.3428
Scaled Pearson X2	2726	3660.3908	1.3428
Log Likelihood		388.6566	

(Contiued)

TABLE 11.5 Model Information for Total Crash Data with Negative Binominal Regression Implemented by SAS PROC GENMOD (Continued)

Algorithm converged.

Analysis Of Parameter Estimates

Parameter	DF	Estimate	Standard Error	Wald 95% Confidence Limits		Chi-Square	Pr > ChiSq	
Intercept	1	1.0921	0.3317	0.4419		1.7423	10.84	0.0010
Right_Lane	1	-0.0798	0.0304	-0.1393		-0.0202	6.89	0.0087
Right_Shoulder	1	-0.0663	0.0067	-0.0794		-0.0532	98.06	<.0001
Dispersion	1	0.4566	0.0378	0.3824		0.5307		

NOTE: The negative binomial dispersion parameter was estimated by maximum likelihood.

Example 11.4

The crash data in Example 11.3 can be refitted using the negative binomial regression model with log(segment length*3) as an offset variable and log(AADT) as one of predictor variables. The functional form for the mean total crashes in that case is give as

E(total crashes for 3 years) =exp{loglength3 + β_0 + β_1 right lane +β_2 right shoulder + β_3 logAADT}

where E(total crashes for 3 years) is the expected number of total crashes for three years and loglength3 = log(segment length*3). Table 11.6 contains the SAS output for the negative binominal regression (with the mean function given above) obtained by using SAS PROC GENMOD as follows:

```
PROC GENMOD DATA=RSBS_2L_NM_LS;
MODEL Total_Crashes_for_3_years = Right_Lane Right_Shoulder
LogAADT
/ DIST=NB LINK=Log OFFSET=LogLength3;
RUN;
```

It can be observed from Table 11.6 that both right lane and right shoulder have a statistically significant influence on total crashes. The negative sign for the coefficients for them suggests that as the width of the right lane or the width of the right shoulder increases, the crash count decreases, which is consistent with expectation. The coefficient for logAADT is estimated to be 1.0654. Because the corresponding confidence interval does not contain 1, we may conclude that the crash count increases slightly faster than AADT increases. The prediction equation obtained by using the estimated coefficients in Table 11.6 can be given as follows:

```
E(Total Crashes for 3 years)
= exp{ Log(Segment Lengthx3) - 7.2208 - 0.0838 Right Lane
-0.0736 Right Shoulder +1.0654 LogAADT}
= 3x(Segment Length) AADT¹·⁰⁶⁵⁴ xexp{-7.2208-0.0838 Right Lane
-0.0736 Right Shoulder }.
```

From the "criteria for assessing goodness-of-fit," it can be seen that the negative binomial model of Table 11.6 fits the data as equally well as the model in Table 11.5. The conclusions on the effects of right lane width or right shoulder width do not change materially regardless of which model form is used. It needs to be noted, however, that there might be other cases where the form of the model makes a difference in both the goodness of fit and the parameter estimates. In such cases, selection of the model can be made based on both the goodness of fit and engineering judgment.

TABLE 11.6   Model Information for Total Crash Data with Negative Binominal Regression with Log(AADT) Included as a Predictor Implemented by SAS PROC GENMOD

The GENMOD Procedure

Model Information

Data Set	WORK.RSBS_2L_NM_LS
Distribution	Negative Binomial
Link Function	Log
Dependent Variable	Total_Crashes_for_3_years Total Crashes for 3 years
Offset Variable	LogLength3 LogLength3

Number of Observations Read	2729
Number of Observations Used	2729

Criteria For Assessing Goodness Of Fit

Criterion	DF	Value	Value/DF
Deviance	2725	2574.5193	0.9448
Scaled Deviance	2725	2574.5193	0.9448
Pearson Chi-Square	2725	3768.3723	1.3829
Scaled Pearson X2	2725	3768.3723	1.3829
Log Likelihood		391.6151	

Algorithm converged.

Analysis Of Parameter Estimates

Parameter	DF	Standard Estimate	Error	Wald 95% Confidence Limits		Chi-Square	Pr >	ChiSq
Intercept	1	-7.2208	0.3721	-7.9501	-6.4916	376.64	<.0001
Right_Lane	1	-0.0838	0.0306	-0.1437	-0.0238	7.49	0.0062
Right_Shoulder	1	-0.0736	0.0074	-0.0880	-0.0592	100.10	<.0001
LogAADT	1	1.0654	0.0270	1.0125	1.1184	1554.42	<.0001
Dispersion	1	0.4509	0.0377	0.3771	0.5247		

NOTE: The negative binomial dispersion parameter was estimated by maximum likelihood.

11.6 CONCLUDING REMARKS

In some transportation applications the dependent variable is discrete. As an example, crash count data (number of crashes per unit of time) and queue length measurements (number of cars waiting at a traffic signal) must have, by definition, nonnegative integer values. In these situations the statistical methods described in Chapter 10 may not be appropriate. Note that this does not mean you *cannot* run the standard regressions techniques with count data—many software packages will allow you to run standard regression models with discrete dependent variables. In this situation, though, the results will often be misleading. As always, it is up to the user to understand the assumptions and limitations of the software he or she chooses to use.

Practicing engineers are often confronted with statistical models related to safety issues. In particular, they will be asked to design or operate the transportation system in the safest manner possible. In order to do, this they will need to know what trade-offs are involved with respect to various control variables, such as lane width, shoulder width, etc. The techniques discussed in this chapter will help them develop models that may provide insight to these relationships.

The use of generalized linear models was the focus of this chapter. These models are ideal when the dependent variable is count data (e.g., integer). The predictor variables can be either discrete or continuous. The two most common models are Poisson and negative binomial regression models, and these are often used in developing crash prediction models, which are known in the literature as safety performance functions (SPFs). These models are easily implemented with existing statistical software such as JMP or SAS. In addition, most state departments of transportation have crash count data along with information on key parameters. As such, most practicing professionals will be exposed to these types of models.

APPENDIX: MAXIMUM LIKELIHOOD ESTIMATION

In Chapter 4 we introduced probability density functions and probability mass functions. For example, the pdf for the normal density is

$$f(x;\mu,\sigma^2)=\frac{1}{\sqrt{2\pi}\sigma}e^{-\frac{(x-\mu)^2}{2\sigma^2}}$$

and the pmf for the Poisson is

$$P(X = x) = \frac{\exp(-\lambda)\lambda^x}{x!}.$$

For each pdf and pmf, the functions are thought of as functions of x that depend upon parameters. In the two cases above, the parameters are (μ, σ) and λ, respectively.

Likelihoods in the simplest form are a pdf or pmf that is thought of as a function of the parameters for fixed values of x. More specifically, let $X_1, \ldots, X_n$ be a sequence of independent and identically distributed random variables. Suppose each random variable has pdf or pmf $f(X_i | \theta)$ for $i = 1, \ldots, n$. The parameter θ may be multidimensional, as in the case of the normal pdf $\theta = (\mu, \sigma)$, or univariate, as in the case of the Poisson $\theta = \lambda$. Then the likelihood function is

$$\prod_{i=1}^{n} f(X_i | \theta).$$

Maximum likelihood estimation maximizes $\prod_{i=1}^{n} f(X_i | \theta)$ with respect to θ, keeping the X s fixed.

The likelihood for the normal distribution after some simplification is

$$L(\mu, \sigma | X_1, \ldots, X_n) = \frac{\exp\left(-\frac{1}{2}\left(\sum_{i=1}^{n} \frac{(X_i - \mu)^2}{\sigma^2}\right)\right)}{(\sqrt{2\pi}\,\sigma)^n}.$$

The likelihood for the Poisson distribution after some simplification is

$$L(\lambda | X_1, \ldots, X_n) = \frac{\exp(-n\lambda)\lambda^{\sum_{i=1}^{n} X_i}}{\prod_{i=1}^{n} X_i!}.$$

Using calculus, it can be shown that for the normal distribution the maximum likelihood estimators for μ and σ are $\bar{X}$ and $\sqrt{(n-1)/n}\,S$. Finally, using calculus, it can be shown that the maximum likelihood estimator for λ is $\bar{X}$. For more detail, see Casella and Berger (1990).

HOMEWORK PROBLEMS

1. Researchers are interested in determining the relationship between crashes on rural two-lane highways and lane and shoulder widths. Use the data set "Crashes_rural 2ln highways." Note that right lane and right shoulder actually represent the widths of them, respectively.

   a. Construct the histogram of the variable "Total crashes for 3 years." What characteristics do you observe from the distribution?

   b. Plot total crashes for three years (Y) vs. right lane (X). Also, plot total crashes for three years (Y) vs. right shoulder (X). Is there any pattern in the plot? Does the relationship between two variables (X and Y) look linear?

   c. Fit the normal linear regression model to total crashes for three years with right lane and right shoulder included as predictors. Do you think the model fits the data well? Create appropriate diagnostic plots. What assumptions seem to be violated?

   d. Fit the normal linear regression model to total crashes for three years with right lane and right shoulder, logAADT, and loglength3 included as predictors. Create appropriate diagnostic plots. What assumptions seem to be violated?

2. Using the same data set ("Crashes_rural 2ln highways"):

   a. Fit a Poisson regression model to the total crashes for three-year data using right lane, right shoulder, and logAADT as covariates, and loglength3 as an offset variable. What are the parameter estimates for right lane, right shoulder, and logAADT and the corresponding standard errors? What effects are statistically significant at $\alpha = 0.1$?

   b. Do you think the Poisson regression model fits the data well? Why or why not?

c. Fit an overdispersed Poisson regression model to the data by using the same set of covariates and an offset variable as in problem 2a. What are the parameter estimates for right lane, right shoulder, and logAADT and the corresponding standard errors? What effects are statistically significant at $\alpha = 0.1$?

d. Fit a negative binomial regression model to the same data using the same set of covariates and an offset variable as in problem 2a. What are the parameter estimates for right lane, right shoulder, and logAADT and the corresponding standard errors? What effects are statistically significant at $\alpha = 0.1$?

e. Do you think the negative binomial regression model fits the data well? Why or why not?

f. What can you conclude about the effects of right lane and right shoulder on total crashes?

g. What about the effect of AADT on crashes? Do you think the relationship between AADT and crashes is linear? Why or why not?

3. Researchers extracted a subset of total crashes that are expected to be influenced by the widths of right lane and right shoulder (such as single vehicle crashes, opposite direction crashes, etc.). In the data set "Crashes_rural 2ln highways," the column named "Related crashes for 3 years" contains such crashes. Answer problems 2a to 2g using "Related crashes for 3 years" instead of "Total crashes for 3 years."

REFERENCES

Boyle, P., F. Flowerdew, and A. Williams. 1997. Evaluating the goodness of fit in models of sparse medical data: A simulation approach. *International Journal of Epidemiology* 26:651–56.

Casella, G. and R. L. Berger. 1990. *Statistical inference*. Pacific Grove, CA: Wadsworth Books./Cole.

Cohen, J. 2003. *Applied multiple regression/correlation analysis for the behavioral sciences*. 3rd ed. Mahwah, NJ: L. Erlbaum Associates.

Fitzpatrick, K., W. H. Schneider, and E. S. Park. 2005. *Comparisons of crashes on rural two-lane and four-lane highways in Texas*. Technical Report 0-4618-1. College Station, TX: Transportation Institute.

Lord, D., S. P. Washington, and J. N. Ivan. 2005. Poisson, Poisson-gamma and zero-inflated regression models of motor vehicle crashes: Balancing statistical fit and theory. *Accident Analysis & Prevention* 37:35–46. http://dx.doi.org/10.1016/j.aap.2004.02.004

McCullagh, P., and J. A. Nelder. 1989. *Generalized linear models*. 2nd ed. London: Chapman & Hall.

Sheather, S. J. 2009. *A modern approach to regression with R*. NewYork, New York: Springer.

CHAPTER 12

Experimental Design

12.1 INTRODUCTION

Much of what we know of transportation phenomena comes from observation. Transportation professionals observe data, build a model, and test their hypotheses. Ultimately, they agree that the model works and can be used for estimation and prediction (within some predefined limits), or they agree that the model is not valid. Of course, a more likely scenario is that they agree that more information is required.

This chapter is focused on designing experiments where data can be obtained to test a driving hypothesis. First, the designed experiment approach is compared and contrasted with the direct observation approach. Subsequently, a discussion of proper experimentation techniques is provided. Lastly, the theory of model development for designed experiments and the methods for testing the underlying hypothesis of the experiment are discussed.

12.2 COMPARISON OF DIRECT OBSERVATION AND DESIGNED EXPERIMENTS

Before the theory of designed experimentation is discussed, it is worthwhile to compare and contrast the two primary methods of obtaining data that transportation engineers will be exposed to during their careers. In transportation engineering the most common methodology is direct observation. For example, a transportation analyst may wish to study gap acceptance during the permitted left turn phases at unsignalized intersections. The analyst would choose one or more representative unsignalized intersections and observe the attributes of the permitted left turning movements (e.g., type of vehicle, size of gaps chosen, speed of oncoming vehicles, size of gaps not chosen, number of lanes, etc.). Based on this information, a gap acceptance model could then be developed.

The second method is based on experimentation in which the analyst controls the environment of the observations. For example, an analyst may set up a permitted left turn experiment on a closed course. The analyst could then vary the gaps between the oncoming vehicles by adjusting, for example, the distance between vehicles, the type of vehicles, their speed, etc. From this information a second type of gap acceptance model can be developed.

There are advantages to both the observational and the experimental approach. The direct observation method has the advantage that, if done correctly, the driver's decision is not affected by the observer. The information obtained is pure in the sense that a driver is making a decision with no outside impacts. One disadvantage is that certain types of data, such as the driver's state of mind, cannot be collected by an outside observer, though it could be important in model development. Similarly, not all of the gaps an analyst may be interested in may be experienced by the driver; that is, it may be too cost prohibitive to conduct the observational study so all combinations of events are observed. Another disadvantage of observational studies is that there are confounding effects (such as location, visibility, etc.) that cannot be adjusted by the modeler. A list of challenges for observational studies is given in Section 12.10.

The advantage of the designed experiment is that many factors can be measured and controlled that cannot be measured in a field study. The downside is that the experiment itself might affect the results. For example, a driver's gap acceptance behavior might be different because the driver knows she is being watched.

For many transportation phenomena the designed experiment is the most economical and efficient method of obtaining data. This is particularly true when human behavior is not involved. It should be noted that sometimes it is impossible to conduct experiments in which only observational data are required. For example, if one were studying the effect of alcohol on safety, it would be unethical to provide alcohol to drivers, let them drive on a public roadway, and then measure the effects on safety. In this situation only indirect observations would be allowed and would take the form of analyzing accident statistics for which alcohol was a contributing factor. Alternatively, alcohol studies can be, and have been, conducted in a safe, closed-course environments, such as test tracks.

Transportation experiments provide an opportunity to develop solutions to important transportation questions. What is the best asphalt concrete mixture for long-term durability? What is the best coating for a stop sign so that it can be seen at night? A well-planned experiment will provide useful information that can help answer these questions. A poorly done experiment can be less than useful by providing inadequate or mistaken information, which results in misleading models.

12.3 MOTIVATION FOR EXPERIMENTATION

To motivate the designed experiment protocols let us consider a situation where a given state transportation agency is interested in studying which of two stop sign coatings, coating A or coating B, is better. For safety reasons, the transportation engineer decides to conduct the experiment on a closed course because an observational study would not be possible for the given hypothesis. The measure of merit for the stop sign coatings is the average distance from the sign that drivers begin to decelerate. Consider the following experiments, which we will label "good" and "poor" for what should be obvious reasons.

12.3.1 The Good Experiment

A sample size calculation using JMP was done to ensure that if drivers decelerate at least 5 feet sooner, on average, for one of the state's commonly used coatings (A or B), the experiment will be 85% likely to detect that there is a difference. Based upon this JMP calculation, one hundred volunteers are randomly chosen. Fifty of them are randomly assigned to drive a closed course that has been outfitted with stop signs treated with coating A. The other fifty drivers drive the course that has been outfitted with coating B. The process by which the number of samples is derived is the focus of this chapter.

The likely outcome from the good experiment is that if one of the stop sign coatings leads to an improvement of an average of 5 feet or more, it is likely that the treatment will be discovered as superior.

12.3.2 The Poor Experiment

It was decided based *a priori* that twelve volunteers would be tested for each sign coating without random assignment, and without using the appropriate statistical formula to calculate the recommended sample sizes.

The likely outcome from the poor experiment is that if one of the stop sign coatings leads to an improvement, it is less likely that the treatment will be discovered as superior. In addition, due to random variation, any difference that is found may indicate that the wrong coating is superior. The experimental outcomes may be dependent upon the attributes of the volunteers. For example, if the experiment were conducted in a university setting, the volunteers might all be undergraduate students in a transportation statistics class. If there were a difference in sign coatings as a function of age of the driver, then this experiment might miss this effect. In summary, the authors would like to stress that a poorly designed experiment will provide worse results, on average, than a properly designed experiment.

Remark 12.1

In most countries experiments involving human and animal subjects require institutional review board (IRB) approval. The motivation for this chapter is based on the assumption that the transportation engineer would be allowed to conduct planned studies. However, as a practical matter, all experimentation involving human subjects will have to go through some type of institutional review board. In addition, transportation agency or research agency lawyers often will be called in to check the protocols. These considerations frequently preclude direct experiments on the transportation network.

On the surface, the proposed study of stop sign coatings would seem reasonable to carry out in a field experiment. Suppose a transportation engineer is allowed to choose between the two sign types with identical costs. Being conscientious, she would like to identify which of the two works better in the field. The designed experiment approach that is the focus of this chapter would allow her to do so; that is, she could locate reasonable stop sign locations, randomly assign signs with different coatings to these locations, and then observe the results. From a legal point of view, however, she might not be allowed to conduct this type of experiment in the field. It would be a decision that would have to be made by an IRB. To illustrate, say coating B is better than coating A (remember, this will not be known until after the experiment). If during the course of the study, someone crashed at one of the coating A sites, which was later found to be an inferior coating type, the transportation agency could be held responsible, even if the crash were unrelated to sign coating. For this reason, these types of designed experiments are rarely done in the field. For the reasons outlined in the above example, most likely the experiments would be conducted on a closed-course environment.

Transportation analysts are often confronted with similar situations: Which of the x allowed for use by the transportation agency, works best in the field and under what conditions? What font size should be used on traffic signs?

The reader often will be confronted with this type of question. That is, the reader will have to choose whether it is best to go to the field and collect observed data (without changing the experiment) or to design an experiment in a closed-course environment. Obviously, the correct choice will be a function of the budget for the study, the underlying hypothesis, past experience, etc. It will also depend upon IRB decisions, as in many countries experiments involving human subjects must get IRB approval before implementation.

12.4 A THREE-FACTOR, TWO LEVELS PER FACTOR EXPERIMENT

Consider the three-factor hypothetical experiment. Suppose we want to assess the effect of background material on a highway sign (two materials: A_1 or A_2), the sign placement (two placements: B_1 or B_2), and the size of the sign (two sizes: C_1 or C_2). We then have three factors, each with two possible levels. Suppose the transportation agency wants traffic to slow down after viewing a particular warning sign. The goal of the research is to identify the best set of factors (e.g., material, placement, and size) that will achieve this objective. Our dependent variable or figure of merit will be measured speed at a specified distance from the sign. We hypothesize that

$$speed = f(A_i, B_j, C_k) + error \qquad (12.1)$$

where $i = 1,2$; $j = 1,2$; $k = 1,2$. The error term in the equation is meant to account for all missing factors, and it is assumed to have a mean of zero and a constant variance. The mean response function can be modeled with the following (over)parameterized form:

$$f(A_i, B_j, C_k) = constant + \alpha_i + \beta_j + \delta_k + (\alpha\beta)_{ij} + (\alpha\delta)_{ik} + (\beta\delta)_{jk} + (\alpha\beta\delta)_{ijk} \qquad (12.2)$$

where $i = 1,2$; $j = 1,2$; $k = 1,2$.

The parameters that have a single subscript, such as α_i and β_j, represent main effects. The parameters with double subscripts, such as $(\alpha\beta)_{ij}$, represent two-way interactions. For example, material (C) and size (A) may have a combined effect that results in a greater (or lesser) effect on speed than their individual effects. Finally, the three-way interaction is represented by $(\alpha\beta\delta)_{ijk}$. Consider the situation where all parameters other than the $\alpha_i,\beta_j,\delta_k$ in the equation are known to be equal to zero. In this case, the value of $f(A_i,B_j,C_k)$ would depend only upon the choice of the main effects (material, placement, and size), and there would be no interaction effects among the factors. On the other hand, if the assumption of no interaction effects was not valid and the parameters with two or more subscripts are not equal to zero, then experiments that change one variable at a time would lead typically to biased models.

12.4.1 A Common Fallacy: Changing One Factor at a Time Experiments

A classical myth about optimal experimental design is that for experiments with many factors it is best to vary only one factor at a time. This myth is dangerous in many ways. The first misconception is that optimizing one variable at a time will give the same result as optimizing all the variables simultaneously. Even if a large number of single variable experiments were practical, in many cases they will not provide a practical optimum. In addition, repeating sequences of single variable experiments costs a lot more to run. If the experiment only changed one variable at a time, then it would be extremely difficult to estimate and understand the role of interactions among the factors. If these interactions were important, the analyst would end up with a model that leads to poor decisions. For example, if size and material had a negative interaction effect, a model that ignored this interaction effect could lead to poor estimation of speed.

Example 12.1

Consider doing a one factor at a time experiment to estimate the parameters of Equation 12.2 as shown in Table 12.1. It can be seen that all the settings were changed one at a time, and there are four runs. By *run* we mean collecting experimental data at the settings indicated by one row of the design table. Equation 12.2 has eight parameters. At an intuitive level, it is clear that four observations would not be enough to estimate the eight parameters needed to characterize the eight possible values for *f*. For example, how would we know

TABLE 12.1   A Design Changing One Factor at a Time for the
Signing Experiment

Run	Level for A	Level for B	Level for C	Value for f
1	A1	B1	C1	f(A1,B1,C1)
2	A2	B1	C1	f(A2,B1,C1)
3	A2	B2	C1	f(A2,B2,C1)
4	A2	B2	C2	f(A2,B2,C2)

the effect of setting the factors to level A2, B1, and C2 (the value f(A2,B1,C2)) if interaction effects are nonzero? By definition, conducting an experiment in this way ignores the interaction effects, which would be problematic if the interaction effects were considerable.

12.5 FACTORIAL EXPERIMENTS

Consider the three-factor signing experiment discussed before. Each factor has two levels and the figure of merit is measured speed. If we are unwilling to assume that any parameters in the model of Equation 12.2 are equal to zero, then we have eight parameters to estimate. Typically constraints are imposed to aid in model interpretation. Without constraints, we would have twenty-seven unconstrained parameters, but only eight possible levels for the function f. The model would be overparameterized using twenty-seven parameters. The constraints imposed in this textbook are consistent with those used by JMP. Other software packages may impose different constraints, so package users need to read the software's user manual in order to properly interpret parameter estimates. The constraints that we impose include:

$$\sum_{i=1}^{2} \alpha_i = 0; \sum_{j=1}^{2} \beta_j = 0; \sum_{k=1}^{2} \delta_k = 0; \sum_{i=1}^{2} (\alpha\beta)_{ij} = 0; \sum_{j=1}^{2} (\alpha\beta)_{ij} = 0;$$

$$\sum_{i=1}^{2} (\alpha\delta)_{ik} = 0; \sum_{k=1}^{2} (\alpha\delta)_{ik} = 0; \sum_{j=1}^{2} (\beta\delta)_{jk} = 0; \sum_{k=1}^{2} (\beta\delta)_{jk} = 0.$$

as well as constraints on the three-way interactions.

The constraints help measure effects from the overall average. For example $\alpha_1 + \alpha_2 = 0$ implies that if $\alpha_1 \neq 0$, then neither is α_2, and one of them is positive and the other negative. Put another way, the effects are measured as deviations from the average.

TABLE 12.2   A Complete Factorial Design for the Signing Experiment with Three Factors, Each Having Two Levels

Background Material	Sign Placement	Sign Size	Speed
A1	B1	C1	
A2	B1	C1	
A2	B2	C2	
A2	B2	C1	
A1	B1	C2	
A1	B2	C1	
A2	B1	C2	
A1	B2	C2	

Given the above constraints, the number of free parameters that need to be estimated is eight. These eight consist of the overall mean (e.g., constant term), the three main effects, the three two-way interactions, and the one three-way interaction. A complete factorial arrangement of treatments is given in Table 12.2 and was generated by the JMP *Design of Experiments* (DOE) module. It can be seen that JMP has identified the eight levels. While this example was small for larger experiments, this option is particularly useful for defining the design possibilities *a priori*.

For this design, all of the main effects and interactions can be estimated because every possible combination of factors is used in the design. It should be noted that collecting the minimum number of observations (e.g., eight in this example) is problematic, as there will not be enough observations to estimate standard errors. Using only eight observations to estimate a model with eight parameters will result in overfitting (e.g., the model fits the data exactly). Consequently, in most studies, additional observations are collected. The extra observations are needed to estimate the model's ninth parameter—the standard deviation of the error. These extra observations typically are done by replicating the experimental runs, using each setting two or more times, or augmenting the experimental design. A methodology for calculating the required number of runs is described later in this chapter. Not surprisingly, randomization is key to a successful designed experiment. For example, the experimental units are usually run in random order. Runs are also usually randomly assigned to geographical location or subjects to reduce bias in experimental outcomes. Experiments with larger numbers of factors involve more complicated constraints (see Kuehl, 2000). By using JMP or another widely used statistical program, the user will get appropriately fitted models and tests.

12.6 FRACTIONAL FACTORIAL EXPERIMENTS

Factorial experiments allow the estimation of the interaction effects of all orders. However, when the number of factors is large, or when there are many possible levels for factors, then the number of experimental runs needed may be very large. For example, if there are seven factors and each factor has two levels, the number of experimental runs needed for a full factorial experiment is $2^7 = 128$. This problem is compounded because standard errors for the estimated parameters often are required in order to estimate confidence intervals of sufficient accuracy. Consequently, even more runs would be required beyond this minimum number. As the reader may be aware, there are few transportation studies that involve this many experiments.

A common solution to this dilemma is to run a preliminary experiment or screening experiment. Screening experiments allow the analyst to identify the important factors under consideration. These factors are then investigated in a more comprehensive analysis. Typically a preliminary experiment might be run to evaluate only the main effects. For example, if an experiment has eleven two-level factors, then only twelve experimental runs would be needed to evaluate the main effects. This may be contrasted with a full factorial experiment that would need $2^{11} = 2048$ experimental runs. It should be noted that in many engineering studies the higher-order interactions are (1) not usually found to be statistically important, or (2) their magnitudes in comparison to the main effects are so small that they can be practically ignored. For these reasons, many transportation engineering models will ignore the higher-order interactions even if they are found to be statistically significant.

Generally fractional factorial experiments, p, for two-level factors are expressed as 2^{p-r} fractional factorials. When $r = 1$, the experiment is considered a half fraction and the numbers of runs needed are half of those from a full factorial. In the example, we presented a 2^{3-1} fractional factorial that used four experimental runs. When $r = 2$ the resulting experimental design is called a quarter fraction. An excellent and classic reference for fractional factorial designs is Box et al. (2005), and another useful reference is the JMP DOE user manual.

After the screening experiment, a fractional factorial experiment, which is a factorial experiment with only a subset of the original parameters, often can be used to obtain good (and sometimes optimal) estimates of the remaining model parameters. Recall the factorial experiment that we presented about speed reduction as a function of high sign type, sign placement, and size of the sign (see Section 12.5). It was shown that eight runs were required in order to have enough data to estimate the eight model parameters. Suppose that the engineer assumed that the three-way interaction parameter—which measured the interaction of the highway sign type, sign placement, and size of the sign—could be considered negligible. This is equivalent to assuming the third-order interactions equal zero (e.g., $\alpha_{ijk} = 0$). In this case, there are only seven remaining unconstrained parameters to estimate. Because all possible orderings of factors are considered in the factorial experiment, any four runs chosen would be a subset of the factorial experiment.

The package JMP provides a subset that allows estimation of the overall mean and the main effects, and also provides an *aliasing* figure that shows the confounding of effects. By aliasing or aliased terms, we mean model terms that cannot be judged independently from each other. More precisely, the magnitude of the effects is redundant. For example, consider modeling observed speeds using the model $Y = \mu_1 + \mu_2 + \varepsilon$. It would not be possible to tell the difference between values of μ_1 and μ_2, and any values that had $\mu_1 + \mu_2 = \mu$ would give the same value for the mean of the observed speeds. We say that μ_1 and μ_2 are aliased or confounded with each other. As a more serious example, the fractional factorial experiment that JMP gives for the signing experiment that we described is shown in Table 12.3.

There are four runs characterized as $A_2B_1C_1$, $A_2B_2C_2$, $A_1B_1C_1$, and $A_1B_1C_2$. Notice that the JMP changes more than one variable at time. For example, in going from setting $A_2B_1C_1$ to $A_2B_2C_2$ the second and third

TABLE 12.3 A Simple Fractional Factorial Design for the Signing Experiment with Three Factors

Background Material	Sign Placement	Sign Size	Speed
A2	B1	C1	
A2	B2	C2	
A1	B2	C1	
A1	B1	C2	

▼ Aliasing of Effects

Effects	Aliases
X1	= X2*X3
X2	= X1*X3
X3	= X1*X2

FIGURE 12.1   Aliasing of effects for the simple fractional factorial design.

factors were both changed. The reader may verify that there is no way to get this design by changing one variable at a time.

Figure 12.1 gives the list of aliased effects. The first line of Figure 12.1, $X1 = X2*X3$, means that the main effect for background is confounded (indistinguishable) from the two-way interaction effect between sign placement and sign size. The other terms are interpreted similarly; for example, the main effect for sign size is confounded with two-way interaction effect between sign background and sign placement. Thus, for example, the design would not let us distinguish between a big effect for sign background and an interaction effect between sign placement and sign size. If these interaction effects were known to be equal to zero, or negligible, then the confounding issue disappears. In this situation a fractional factorial experiment would be appropriate—with a corresponding reduction in data and expense.

12.7 SCREENING DESIGNS

Fractional factorial designs are a subset of screening designs. Screening designs can be used to find the factors that likely will lead to the most useful model. Screening designs allow for the investigation of a large number of factors (usually $p \leq 100$). For example, a screening design for eleven factors needs only twelve experimental runs, whereas a screening design for fifteen factors requires only sixteen experimental runs, and a screening design for thirty-one factors requires only thirty-two experimental runs.

These are called Plackett–Burman designs. Recall that full factorial designs would require $2^{11}, 2^{15}$ and 2^{31} experimental runs, while a half fractional factorial experiment would require half as many runs, etc. These designs can be obtained in JMP under the DOE menu.

The authors suggest that the JMP aliasing figure be reviewed before implementing a screening design. For example, a screening fractional factorial design for fifteen factors can be obtained from JMP that produces

▼ **Aliasing of Effects**

Effects	Aliases
X1	= X2*X15 = X3*X14 = X4*X11 = X5*X6 = X7*X8 = X9*X10 = X12*X13
X2	= X1*X15 = X3*X13 = X4*X10 = X5*X7 = X6*X8 = X9*X11 = X12*X14
X3	= X1*X14 = X2*X13 = X4*X8 = X5*X9 = X6*X10 = X7*X11 = X12*X15
X4	= X1*X11 = X2*X10 = X3*X8 = X5*X12 = X6*X13 = X7*X14 = X9*X15
X5	= X1*X6 = X2*X7 = X3*X9 = X4*X12 = X8*X15 = X10*X14 = X11*X13
X6	= X1*X5 = X2*X8 = X3*X10 = X4*X13 = X7*X15 = X9*X14 = X11*X12
X7	= X1*X8 = X2*X5 = X3*X11 = X4*X14 = X6*X15 = X9*X13 = X10*X12
X8	= X1*X7 = X2*X6 = X3*X4 = X5*X15 = X9*X12 = X10*X13 = X11*X14
X9	= X1*X10 = X2*X11 = X3*X5 = X4*X15 = X6*X14 = X7*X13 = X8*X12
X10	= X1*X9 = X2*X4 = X3*X6 = X5*X14 = X7*X12 = X8*X13 = X11*X15
X11	= X1*X4 = X2*X9 = X3*X7 = X5*X13 = X6*X12 = X8*X14 = X10*X15
X12	= X1*X13 = X2*X14 = X3*X15 = X4*X5 = X6*X11 = X7*X10 = X8*X9
X13	= X1*X12 = X2*X3 = X4*X6 = X5*X11 = X7*X9 = X8*X10 = X14*X15
X14	= X1*X3 = X2*X12 = X4*X7 = X5*X10 = X6*X9 = X8*X11 = X13*X15
X15	= X1*X2 = X3*X12 = X4*X9 = X5*X8 = X6*X7 = X10*X11 = X13*X14

FIGURE 12.2 Aliases for a fifteen-factor screening design assuming third- and higher-order interactions are zero.

the aliasing shown in Figure 12.2. It can be seen that effect X1 cannot be distinguished from eight interaction terms in the full model (e.g., X2-X15, X3-X14, X4-X11, X5-X6, X7-X8, X9-X10, X12-X13). The reader should immediately see the trade-offs involved. There is a significant reduction in the amount of data required. However, the modeler is giving up the possibility of analyzing various interaction effects, as shown in Figure 12.2. Some interactions will clearly be zero. For example, if X2 is vehicle color and X15 is lane width, their interaction term would probably be zero. Obviously, each experiment will be different and the user will have to decide how comfortable she is in making these trade-offs.

Figure 12.2 shows that the main effects are confounded with various second-order interactions, but if all interaction terms were zero, then the main effects could be estimated without bias. When screening designs are run, it is hoped that the interaction terms are much less important than the main effects. It is not usually practical to use a fractional factorial design 2^{p-1} or 2^{p-2} when the number of factors, p, is large, say, 15.

Screening designs can be used in microsimulation experiments, where road design elements, driver behavior, and supply and demand are all factors under study.

12.8 D-OPTIMAL AND I-OPTIMAL DESIGNS

When transportation engineers are confident in the form of their model, as described in Chapters 8 and 10, experimental designs can be optimized. To illustrate, consider an asphalt pavement sample where we want to model surface roughness, as measured by international roughness index (IRI) as a function of age (pavement age in years), and traffic volume, as represented by the average annual daily traffic (AADT). We might use a model like this:

$$IRI = \beta_0 + \beta_1 * age + \beta_2 * AADT + \varepsilon.$$

Our goal is to estimate the model parameters, β, and their uncertainty. If we have a choice of pavement ages and AADT, which parameter values should we choose? Similarly, readers may have the need to design experiments (including microsimulation experiments) that use models that are more complex. As long as the model is linear in the βs, we can use the custom design feature in JMP to give an optimal design.

Generally, the transportation engineer will input to JMP the numbers of dependent variables (IRI, rutting, etc.) and the numbers of independent (predictor) variables. He or she also needs to tell JMP which variables are continuous, categorical, and bounds on these variables. In addition, JMP can handle linear constraints on the independent variables.

Let us consider a more complex version of the roughness model that was introduced:

$$IRI = \beta_0 + \beta_1 * age + \beta_2 * age^2 + \beta_3 * AADT + \beta_4 * AADT^2$$
$$+ \beta_5 * age * AADT + \varepsilon.$$

Let us suppose that pavement age varies between one and seven years old, while AADT varies between 50,000 and 1 million vehicles.

The optimal design from JMP for this model is to use the eight runs in Table 12.4. The locations for this design should, if possible, be chosen randomly from those available that meet the preselected criteria. By using this design, roughness needs to be measured at only a few locations in order to develop a meaningful model.

The two optimality criteria that are available in JMP are D- and I-optimality. D-optimality chooses a design that minimizes the generalized variance of the βs. In essence, it minimizes the area of the confidence ellipse for the βs. I-optimality minimizes the average prediction variance. The average is taken over the range of predictors given to the design. Often the two designs coincide, but not always.

TABLE 12.4   Optimal
Design from JMP for
Complex IRI Model

Age	AADT
7	525,000
7	50,000
4	1,000,000
4	525,000
1	525,000
1	1,000,000
1	50,000
7	1,000,000

TABLE 12.5   Optimal
Design for Simple Model

Age	AADT
7	1,000,000
1	50,000
1	1,000,000
7	50,000

A defect of optimal designs is that they are optimal for the specified model, but if the model is wrongly specified, the design can be bad. For example, the optimal design for the original pavement model $IRI = \beta_0 + \beta_1{}^* age + \beta_2{}^* AADT + \varepsilon$ has only four runs and no possibility of identifying quadratic or cross-product terms, as shown in Table 12.5.

Generally, it is wise to choose a design for a more complex model than one believes holds. This model should reduce to the more trusted model if the extra parameters are indeed zero. Good references for optimal designs are Mitchell (1974) and Atkinson and Donev (1992).

D-optimal designs have been used in transportation research, such as those used to assess pavement performance. In Freeman et al. (2006), a D-optimal design was generated for a twenty-three-factor experiment to assess asphalt cement properties. The number of runs that would be needed for a factorial experiment is $2^{23} = 8,388,608$. The D-optimal design generated for estimating main effects and two-way interactions required only 277 runs. Because each run required thirty to forty-five minutes, the savings permitted by a D-optimal design are apparent; in fact, a full factorial experiment would not have been possible.

The DOE module in JMP has a powerful custom experimental design capability. It can be used when the experimenter is reasonably sure of the important factors needed to model in the experiment. Alternatively, when the needed number of factors is small, a custom design also may be sought. Continuous, categorical factors can be used in the design as well as many other types of factors, such as covariates. The JMP custom design module is a very powerful tool that can provide experimental designs for many situations faced by transportation professionals. It will generate designs for any linear model that the engineer provides. In order to use the custom design module, the transportation professional needs to know how many experimental runs that he or she can afford, as the custom design gives choices from a minimum design that simply allows estimation of the basic model parameters, to a user-specified number of experimental runs. It should be noted that it is often useful to do preliminary screening design runs of the factorial model before any data are collected, so that the user will be allowed to make rationale data collection decisions (e.g., what variables to collect, how many should be collected) with regard to the study budget. The reader is referred to the JMP *Design of Experiments* manual for more detail.

12.9 SAMPLE SIZE DETERMINATION

In some cases, it is possible to specify sample sizes based upon type I error, power, a plausible alternative hypothesis, and standard error estimate. The JMP module that helps with these calculations is given within the DOE module, and it has a radio button labeled "Sample Size and Power." Selecting that button produces Figure 12.3.

While the JMP DOE manual has detailed instruction for using this module, we will demonstrate how it can be used to find appropriate sample sizes to compare the effect of two sign backgrounds. Suppose that the measure of effectiveness is measured by the change in speed (measured in MPH), and that a 3 mph change in speed is considered important to detect. Suppose that the standard deviation for mph measured using double-loop detectors is 4 mph. In addition, suppose that we choose a type I error size of .05 and want at least 85% power to detect a difference in mph of at least 3 mph. Then we choose the "Two Sample Means" radio button from the sample size menu and fill it in as shown in Figure 12.4.

We then choose the "Continue" button. The sample size suggested is 65.8 or 66 total observations, or 33 for each treatment. As well, it assumes

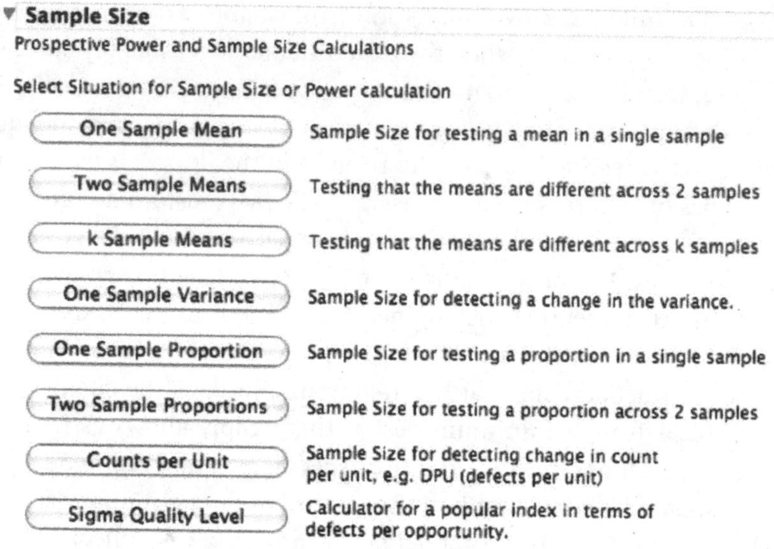

▼ Sample Size

Prospective Power and Sample Size Calculations

Select Situation for Sample Size or Power calculation

(One Sample Mean)	Sample Size for testing a mean in a single sample
(Two Sample Means)	Testing that the means are different across 2 samples
(k Sample Means)	Testing that the means are different across k samples
(One Sample Variance)	Sample Size for detecting a change in the variance.
(One Sample Proportion)	Sample Size for testing a proportion in a single sample
(Two Sample Proportions)	Sample Size for testing a proportion across 2 samples
(Counts per Unit)	Sample Size for detecting change in count per unit, e.g. DPU (defects per unit)
(Sigma Quality Level)	Calculator for a popular index in terms of defects per opportunity.

FIGURE 12.3   JMP sample size menu.

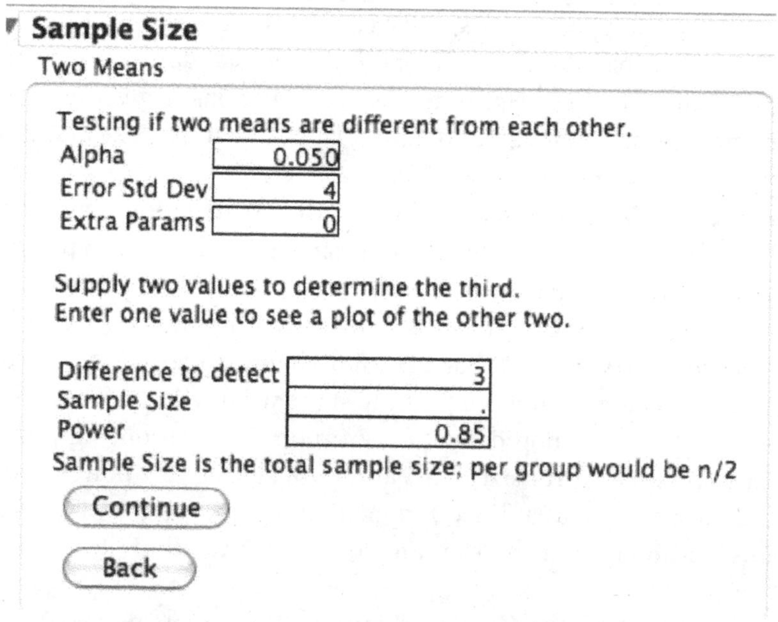

▼ Sample Size

Two Means

Testing if two means are different from each other.

Alpha   0.050
Error Std Dev   4
Extra Params   0

Supply two values to determine the third.
Enter one value to see a plot of the other two.

Difference to detect   3
Sample Size   .
Power   0.85

Sample Size is the total sample size; per group would be n/2

(Continue)

(Back)

FIGURE 12.4   Sample size calculation menu for two sample means.

the analyst does not know the mean but does know the standard deviation (perhaps from previous experience). Consequently, while these sample size tools are very useful, it is up to the analyst to understand their limitation while using them. If the input information is incorrect, it is unlikely that the number of samples will be correct.

If we do not wish to specify power ahead of time, the power can be left unspecified and selecting "Continue" from Figure 12.4 produces Figure 12.5.

Using this figure, we can see how the need for larger sample sizes varies with requirements for larger power. If we use fifty (independent) vehicles for each sign background and thus a total sample size of a hundred, we are very likely able to detect a difference of three mph. As an aside, the reader should be used to seeing these types of relationships by now. In general, the larger the sample size, the better the result; however, the benefit to an increasing sample size decreases at an increasing rate. This is very typical in statistics, which is why it is important to understand the costs of each additional experiment and the benefit associated with having a better answer.

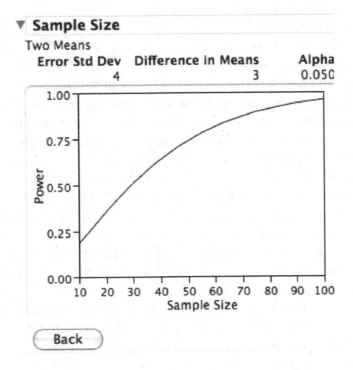

FIGURE 12.5   Power curve.

12.10 FIELD AND QUASI-EXPERIMENTS

As stated in Section 12.2, a transportation professional often has to conduct a field study to assess the impact of different strategies. Generally, observational studies require larger sample sizes than designed experiments. Unfortunately, they also have many uncontrolled factors, as discussed in Chapter 10. In addition to performing regression discontinuity designs, and running ANOVAs, it is important to understand the confounding factors. The following discussions are a summary of material from Campbell and Stanley (1963), Cook and Campbell (1979), and Shadish et al. (2002).

In order to motivate our approach, consider the challenges of modeling policy changes, such as an increase in the energy tax and its effect on average daily traffic (ADT), an increase in a speed limit and its effect on fatalities, or a change in the minimum age for consumption of alcoholic beverages and its effect on crashes. Clearly, a controlled randomized experiment cannot be implemented to judge the effects of these types of possible policy changes. As such, we need to know a good way to analyze this type of program after it has been implemented and what the limitations are of observational studies.

There are two types of challenges to interpreting observational studies. The first are those internal to the experimental subjects. These challenges result in potential misinterpretation of the social change effects for the experimental subjects. The second types of challenges are those external to the experimental subjects. These make it hard to have confidence that findings about the effect of social changes for the experimental subjects are likely to occur for the general population. External challenges have a lot to do with analyzed samples that lack typical population traits. For example, it would be hard to generalize driver responses to sign changes in rural Texas to New York City. It would also be hard to generalize observations in passive work zone enforcement areas (just flashing police lights) to active work zone enforcement areas (tickets given to violators).

Table 12.6 lists the main challenges to internal validity for the analysis of a social program. It is based upon Campbell and Stanley (1963) and Table 2.4 in Shadish et al. (2002).

It is important to note that through sophisticated statistical techniques, some of the challenges may be somewhat mitigated. A classic reference for before and after studies is Hauer (1997). However, the authors emphasize that these challenges will never be ignorable. For example,

TABLE 12.6   Threats to Internal Validity

Challenge or Threat to Internal Validity	Example
1. History. The specific history occurring between before and after measurements.	Cars have improved safety features blunting the effect of increased speed limits.
2. Maturation. Changes in subjects that naturally occur independently from the social change being evaluated.	Measuring the effect of alcohol on young adults in a multiyear study. As people get older, they learn how to lessen the effects of alcohol.
3. Testing. Subjects who do not act as they normally would because they know that they are being evaluated.	The presence of someone with a radar gun tends to slow drivers down.
4. Changes in instrumentation or definitions used to record outcomes.	Changes in how accidents are recorded can lead to difficulty in assessing accident trends.
5. Regression. Selecting matched control and exposed groups for comparison based upon unusual outcomes. These outcomes likely are not repeatable, and the matched groups revert to unmatched levels independently of the social change being evaluated.	Regions are matched based upon ADT for a short period of time. Unusually high ADT will tend to revert to the norm independently of traffic management changes.
6. Biased selection of comparison groups.	Drivers in Texas and New York have different vehicle characteristics. Evaluations must take these differences into account.
7. Biased departure from participation in the evaluation. Remaining subjects may not represent typical outcomes.	Regions where red light cameras are unsuccessful may discontinue their use. The remaining regions using red light cameras will show success.
8. Ambiguous timing of social change implementation.	There is a lag from the time speed limits are changed by municipalities and the time that street signs are posted. Before and after studies need the date that the posted speed limits changed for each road segment.

Bayesian techniques can be used to model and assign a variance to the implementation times for social change. This will mitigate, but not solve, the problem of ambiguous timing for social change. After giving the analogous table (Table 12.7) for challenges to extrapolating predictions to nonexperimental units, a few common quasi-experimental designs will be evaluated. Discussions that are more extensive are found in the cited references.

TABLE 12.7   Threats to External Validity

Challenge or Threat to External Validity	Example
1. Reactive effect of testing	Probe-vehicle drivers, knowing that they are monitored by GPS, drive differently during data collection than they would if they did not know that they were being monitored.
2. Interaction effects of selection biases	Selecting volunteering cities to participate in signing experiments may lead to cities being selected that are not typical.
3. Multiple program effects	Using the same set of experimental units for various social changes may produce results dependent upon all the prior social changes to which the subjects were exposed.

We discuss a few designs from the references to emphasize the importance of quasi-experimental design choices. Let X denote the implementation of a social program, let O denote an observation or measurement, and let R denote randomized assignment to experimental groups among the subjects at hand.

The simplest and least effective design that we consider is the one-shot case study. It is denoted as

$$X \qquad O.$$

There is no control group, and the main point of introducing this design is to have a bad actor to reference. The one-shot case study has to compare the after-program analysis to our opinion of what would have happened if the social change were not implemented. The analysis from this design suffers from all the internal and external challenges that we discussed previously.

The next design is the one-group pretest/posttest design. It is symbolized by

$$O \qquad X \qquad O.$$

Here there is no assignment; all subjects are observed before and after implementation of the intervention. As an example, consider observing average speeds before and after statewide speed limits are lowered from 70 mph to 55 mph. This design does slightly better than the first, but it is far from a good design. For example, if we were testing the effectiveness of banning handheld cell phones over the course of a few years, we would have many other confounding factors. Changing speed limits, vehicle and road design changes, weather differences, and traffic volume changes all would provide challenges to the analysis. At least for challenges to internal validity, this design provides protection for regression and selection biases. See Campbell and Stanley (1963) for more detail.

The final design that we consider is the pretest-posttest control group. It is symbolized by

$$
\begin{array}{cccc}
R & O & X & O \\
R & O & & O.
\end{array}
$$

The subjects available (perhaps all volunteers) are randomly assigned to treatment and control (nontreatment) groups. Both groups are observed before and after implementation of the program. This design mitigates most of the threats to internal validity presented, but due to the fact that the pool of experimental subjects was not chosen at random, little help is provided for extrapolating results found among experimental units to a broader population.

A complete description of these approaches is beyond the scope of this textbook. However, most transportation engineers are confronted with analyzing policy decisions (e.g., what is the effect of raising the speed limit?), and we wanted to introduce the common techniques and, more importantly, discuss the challenges of conducting these types of experiments. A more complete and extensive discussion for the science of quasi-experimental design can be found in the following cited references: Campbell and Stanley (1963), Cook and Campbell (1979), and Shadish et al. (2002).

Remark 12.2

In practice, the assessment of the impact of policy changes in transportation, such as raising the speed limit, is often carried out by observational studies due to the limitation in randomization (random assignment between treatment and nontreatment groups is usually not possible in safety studies). In the cases where the treatment and nontreatment groups are not randomly assigned, the nontreatment group is referred to as a comparison group rather than a control group' Interested readers are referred to Hauer (1997) for more discussions on this issue.

Remark 12.3

We note that there are some semantic differences among terms used in engineering literature (e.g., Hauer, 1997) and related social science literature (e.g., Campbell and Stanley, 1963; Cook and Campbell, 1979; Shadish et al., 2002) in bias assessment. For example, Hauer (1997) considers regression to the mean as a selection bias, whereas the other researchers handle regression to the mean as a separate bias involved in selection. The differences in the way these sources treat these bias terms is essentially semantic; all are deeply concerned with bias issues.

12.11 CONCLUDING REMARKS

This chapter dealt with an important topic—the use of designed experiments. For some specialties, such as safety studies on the effects of alcohol, designed field experiments are out of the question. In other areas they are common, such as sign legibility studies conducted in a closed course, or pavements and material thickness vs. wear studies. When they are able to be used, designed experiments are very efficient at identifying important relationships. Their power comes from the fact that the designer can choose the levels of factors that he or she would like to know more about or change. Consequently, the number of observations can be significantly lower than those of an observational study.

One of the prime benefits of factorial designs is that interaction effects can be explicitly modeled and estimated. For example, in pavement design it is a known fact that while both traffic loading and environment affect pavement life, their interaction term can be important. Note that, as always in statistics, there is an art as well as a science to building models. Sometimes the transportation analyst judges or even intuitively assumes that certain interaction terms are unimportant, possibly recognizing, for example, that there is no reason for a significant interaction (e.g., traffic loading and sign placement on pavement life) or because the magnitude of the interaction is such that it can reasonably be ignored. (In practice, it is uncommon for interactions above three-way interaction to be important.) In these situations, the analyst can reduce the number of runs, which would reduce cost or increase the number of parameters that can be examined for the same cost.

Lastly, it is crucial that the modeling and inference be done within the bounds of good statistical theory and practice. Opinion should be verified with data and valid analyses; this is not a new concept and has been an ongoing theme throughout the textbook. This chapter lists a number of threats to both internal and external validity of the models. It is important that these be addressed before any experimentation begins. For sophisticated model development, the authors recommend the transportation analyst work closely with trained statisticians. While this may result in added expense, it will be far cheaper than a failed experiment or, worse, a poorly defined model that results in erroneous decisions.

APPENDIX: CHOICE MODELING OF EXPERIMENTS

Choice Modeling

Transportation engineers use surveys to provide valuable information for transportation decisions. What are the most preferred modes of travel and under what conditions? Where is the best location to build a shopping mall? Crucial decisions cannot be made without information from important stakeholders. In this section, we will show how to use JMP to design and analyze a survey.

Conducting a survey may require IRB approval. One of the first steps in planning a survey is to get appropriate approvals. The next step is to create clear surveys and have an analysis strategy that supports the decisions that need to be made. For example, consumers may have choices in how they travel, whether by private car, rail, bus, or taxi. They have different trip purposes: work, shopping, or leisure. They also have a choice to make about the time of day to travel.

A properly designed survey and analysis will provide useful information. Both designing and analyzing the survey are necessary to ensure that well-informed decisions are made. There is a choice design module in JMP, and in order to use it, some software-specific vocabulary is required. Note that "Factors" represent types of choices. For example, mode of travel—private car, rail, bus, or taxi—represents a factor with four levels.

A profile is a set of factors at particular levels. An example of a profile is private car, work trip, at 7:30 a.m. The factors are mode of travel, reason for a trip, and time of a trip.

A choice set is a collection of profiles that a person is asked to choose from. For example, a person may be asked what is the most common trip that he takes: (1) private car, work trip, at 7:30 a.m., or (2) bus, shopping trip, at 10 a.m. A person typically would be asked to choose from two to four profiles for each of perhaps ten choice sets on a survey.

An example survey generated by JMP is shown in Figure 12.6.

	Choice Set	Choice ID	Mode	Purpose	Time of Trip
1	1	1	Rail	Work	Mid-afternoon
2	1	2	Car	Shopping	Mid-morning
3	2	1	Bus	Shopping	Noon
4	2	2	Car	Leisure	Mid-morning
5	3	1	Bus	Leisure	Mid-morning rush hour
6	3	2	Rail	Shopping	Mid-morning
7	4	1	Rail	Leisure	Noon
8	4	2	Car	Work	Morning rush hour
9	5	1	Rail	Shopping	Morning rush hour
10	5	2	Car	Work	Noon
11	6	1	Bus	Shopping	Mid-morning
12	6	2	Rail	Work	Noon
13	7	1	Bus	Work	Mid-morning
14	7	2	Car	Shopping	Morning rush hour
15	8	1	Bus	Shopping	Morning rush hour
16	8	2	Taxi	Shopping	Mid-afternoon
17	9	1	Taxi	Work	Morning rush hour
18	9	2	Car	Work	Evening rush hour
19	10	1	Car	Work	Mid-afternoon
20	10	2	Taxi	Work	Evening rush hour

FIGURE 12.6   An example of a choice survey designed by JMP.

Once the responses from the survey are collected, a model is fit to the data. The JMP help file gives the form in which the data must be placed. The model that is fit is the multinomial logit model; see Washington et al. (2003). That model is described by the equations

$$p_{js}\left(\begin{array}{c}\text{option } j \text{ chosen} \\ \text{in choice set } s\end{array}\right) = \frac{\text{EXP}(\beta_j X_{js})}{\sum\limits_{i=1}^{J} \text{EXP}(\beta_i X_{is})}.$$

HOMEWORK PROBLEMS

1. From the published literature, find a field study. List the internal and external threats to the conclusions that the authors present.

2. For the study that you used for question one, what are the factors? Pretend that you can perform a designed experiment for this study.

How many observations would you use and what design would you use? Explain.

3. Generate a 2^{5-2} fractional factorial experiment. What effects are confounded (aliased) with the main effects?

REFERENCES

Atkinson, A. C., and A. V. Donev. 1992. *Optimal experimental designs*. Oxford: Clarendon Press.

Box, G. E. 2005. *Statistics fo experimenters: Design, innovation, and discovery*, 2nd ed. New York: Wiley.

Campbell, D. T., and J. C. Stanley. 1963. Experimental and quasi-experimental designs for research on teaching. In *Handbook of research on teaching*, ed. N. L. Gage. Chicago: Rand McNally.

Cook, T. D., and D. T. Campbell. 1979. *Quasi-experimentation: Design and analysis for field settings*. Chicago: Rand McNally.

Freeman, T. J., J. Uzan, D. G. Zollinger, and E. S. Park. 2006. *Sensitivity analysis of and strategic plan development for the implementation of the M-e design guide in TxDOT operations*. Vols. 1 and 2, Technical Report.0-4714-1. College Station, TX: Texas Transportation Institute.

Hauer, E. 1997. *Observational before-after studies in road safety*. Oxford: Pergamon Press.

Kuehl, R. O. 2000. *Design of experiments: Statistical principles of research design and analysis*. Duxbury, MA: Pacific Grove.

Mitchell, T. J. 1974. An algorithm for the construction of D-optimal experimental designs. *Technometrics* 16:203–10.

Shadish, W. R., T. D. Cook, and D. T. Campbell. 2002. *Experimental and quasi-experimental designs for generalized causal inference*. Boston: Houghton Mifflin.

Washington, S. P., M. G. Karlaftis, and F. L. Mannering. 2003. *Statistical and econometric methods for transportation data analysis*. Boca Raton, FL: Chapman & Hall/CRC Press.

CHAPTER 13

Cross-Validation, Jackknife, and Bootstrap Methods for Obtaining Standard Errors

13.1 INTRODUCTION

A focus of this textbook has been on statistical modeling and estimation—whether as a single variable (e.g., speeds at a location) or as multiple variables (e.g., relationship between traffic speed and weather) using multiple linear regression or ANOVA. As stated before, readers should not only be well versed in developing models and using them for estimation and prediction, but they should also be able to quantify statistically how good their resulting estimates are. To answer this question, the reader has to understand the concept of standard error; consequently, every chapter that deals with inference has a section on estimating standard errors. These techniques, while very powerful, are limited by the underlying assumptions related to a closed-form formula. The use of coefficients of variation, logs, exponents, and other nonlinear functions of the data is commonplace. For example, in transportation engineering origin-destination (OD) estimates and 85th percentile speeds do not have closed-form standard error formulas. In general, nonlinear estimators do not have closed-form formulas for their standard errors. This chapter focuses on three approaches for quantifying uncertainty when a closed-form formula for standard errors is not available. Sometimes propagation of error (see Chapter 4) can be used, but the methods given here are often more accurate for nonlinear estimators. This chapter will focus on approaches that can be employed when you encounter this type of situation.

13.2 METHODS FOR STANDARD ERROR ESTIMATION WHEN A CLOSED-FORM FORMULA IS NOT AVAILABLE

Many statistical estimators given in this textbook—such as the median (or other percentiles) and transformations such as $\ln(\overline{X})$ and R^2 —do not have simple formulas for obtaining standard errors. In these situations other techniques are required. For example, in Chapter 4 we illustrated propagation of error techniques for obtaining approximate standard errors for nonlinear transformations. For many decades, propagation was the most practical way to get standard errors for nonlinear methods. However, approaches developed over the past fifty or so years are often better because they have been shown to provide more realistic uncertainty estimates. The main defect of propagation of error is that this approach will often ignore the higher-order terms from the Taylor expansion. In many instances, these terms contribute to uncertainty, but if they are ignored, the resulting standard error (SE) estimates can be underestimated.

The three leading modern approaches to standard error estimation are the cross-validation, bootstrap, and jackknife methods. Each of these approaches has several ways in which it can be implemented. To initiate this discussion, consider Figure 13.1, which shows measured travel time on a

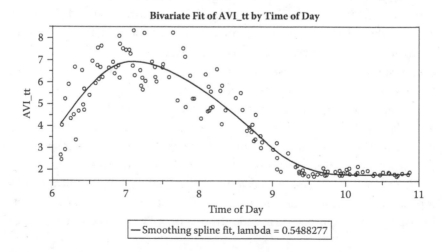

FIGURE 13.1   An example of a smoothed nonlinear estimator.

corridor in Houston, as quantified by the Automatic Vehicle Identification system, vs. the time of day. The data are fully described in Eisele (2001) as well as in Eisele and Rilett (2001). A fitted smoothing spline (denoted by a solid-line curve) was used to estimate the relationship between the travel time and the time of day. See Green and Silverman (1994) for more information about splines and other nonparametric estimators. Note that the fitted smoothing spline does not have a simple closed-form standard error formula. In this situation, engineers who wish to calculate the SE, and hence the CI or PI, will need to use alternative methods for finding practical uncertainty estimates. This will be the focus of this chapter, though, unfortunately, space limitations restrict a full conversation. Likewise, the theoretical underpinnings are beyond the scope of this book. Readers who want a more in-depth treatment are referred to excellent books written by Efron and Tibshirani (1993) and Mosteller and Tukey (1977).

13.3 CROSS-VALIDATION

Cross-validation's main use is to provide uncertainty estimates for predictions as contrasted from parameter estimates. They are used to measure how well we can predict a new Y. In Chapter 10, the error variance estimate (i.e., prediction error estimate) or MSE was given as

$$
\hat{\sigma}^2 = \frac{\sum_{i=1}^{n}\left(Y_i - \hat{\beta}_0 - \hat{\beta}_1 X_{1i} - \ldots - \hat{\beta}_p X_{pi}\right)^2}{n-p} = \frac{\sum_{i-1}^{n}\left(Y_i - \hat{Y}_i\right)^2}{n-p}.
$$

The degrees of freedom, $n-p$, is the proper denominator based upon statistical calculations. The degrees of freedom calculations depend upon having a linear estimator and a finite number of parameters. For nonlinear parametric models propagation of error techniques often can be used to obtain an equivalent estimator of prediction error. For estimators that do not allow a Taylor expansion or have the number of parameters increasing with sample size—such as a smoothing spline—other approaches are needed. For prediction, a commonly used method is cross-validation. The most common form of cross-validation is called leave-one-out cross-validation. Let the smoothing spline estimate in Figure 13.1 be denoted as $\hat{f}(x)$. It was obtained using all the data plotted. Suppose we denote these data as $(x_i, Y), i = 1, 2, \ldots n$. Let the smoothing spline estimator that uses all the data *except* the jth point (x_j, Y_j) be denoted as $\hat{f}_{-j}(x)$. Then the cross-validated estimate of prediction error is

$$\hat{\sigma}^2_{cv} = \frac{\sum_{j=1}^{n} \left(Y_j - \hat{f}_{-j}(x_j)\right)^2}{n}. \tag{13.1}$$

In Equation 13.1 the cross-validated estimate of the prediction errors is simply an average of estimated squared errors. The estimated errors are obtained by subtracting the best estimate of the mean function for Y_j using all the observations that are independent of Y_j (i.e., $\hat{f}_{-j}(x_j)$).

In large samples, this technique furnishes a reasonable estimate of the prediction error provided the observations are independent or very close to independent. When the data are dependent, other forms of cross-validation are better suited for prediction error estimation. Readers are referred to Efron and Tibshirani (1993).

13.4 THE JACKKNIFE METHOD FOR OBTAINING STANDARD ERRORS

Another alternative to the propagation of error formulas is the jackknife method. It can provide a standard error estimate for almost any estimator. Both the jackknife and cross-validation procedures use estimates constructed from data sets that use all but one observation. Each observation is deleted once. Unlike cross-validation that is used to estimate prediction mean squared error, the jackknife provides standard errors for parameter estimates. The difference is akin to the distinction between prediction intervals and confidence intervals for regression at a fixed value of x (refer to Chapter 10 for a full discussion).

One approach to the jackknife is based upon the Quenouille-Tukey jackknife: see Tukey (1958). It is mainly used for small- to medium-size samples and where Taylor expansions are difficult to apply. Notwithstanding, propagation of error works well in large samples with independent data, and the presence of a suitable Taylor with small second- and higher-order terms. For linear estimators the jackknife is unnecessary because correct and simple closed-form formulas are available.

The jackknife method is based on pseudovalues. We assume that we have n independent samples. We denote our nonlinear estimator as $\hat{\Theta}_n$, with the subscript emphasizing the number of samples that are used. Let the estimators that use all but the jth sample be denoted by $\hat{\Theta}_{n-1,-j}$. Then the jth pseudovalue is

$$Y_j^* = n\hat{\Theta}_n - (n-1)\hat{\Theta}_{n-1,-j}. \tag{13.2}$$

As an example, let $\bar{Y}^* = \sum_{j=1}^{n} Y_j^* / n$. Then the jackknife estimate for the standard error of $\hat{\Theta}_n$ is

$$\sigma(\hat{\Theta}_n) = \sqrt{\frac{\sum_{j=1}^{n}(Y_j^* - \bar{Y}^*)^2}{n(n-1)}}. \tag{13.3}$$

Note that this approach essentially calculates a series of pseudovalues by removing a single observation from the sample and executing this n times (where n is the number of samples). These jackknife pseudovalues are then centered and used to estimate the SE. In this case the traditional estimator for standard error is obtained. This approach is best illustrated by way of examples.

Example 13.1

We first use a linear example. Please note that using the jackknife approach on a linear example is unnecessary, but it is useful for demonstrating the general methodology. Assume we are conducting a speed study and have recorded speeds from three vehicles of 61, 64, and 68 mph. The mean speed is 64.33 mph. Using the usual formula for standard error of the mean, we obtain a result of 2.03 mph. When we construct the pseudovalues using Equation 13.2, we find that they are 61, 64, and 68 mph, respectively. Thus, the standard error obtained using Equation 13.3 is 2.03 mph.

Example 13.2

Suppose that researchers want to get the standard error for the natural log of the mean of observed speeds. That is, the estimator of interest is

$$Y = \ln(\bar{X})$$

where $\bar{X}$ denotes the average speed. In this case, identifying the standard error for the natural log of the mean, $\ln(\bar{X})$, is considerably more complicated than the previous example. For the data in Example 13.1 it can be easily calculated that the mean of the natural log of the data is 4.16. The pseudovalues for this model are 4.1100, 4.1559, and 4.2189 mph, respectively. The jackknife standard error is .0316 mph. When we use the propagation of error formula we have $\sqrt{S_{\bar{X}}^2 / \bar{X}^2} = .0315$.

As discussed in the literature, the jackknife formula is more accurate than the propagation of error formula (see Efron and Tibshirani, 1993). In the authors' experience, the jackknife standard error is also considerably easier to use than deriving the propagation of error formula. The jackknife standard error has degrees of freedom approximately equal to the number of unique pseudovalues minus 1. This number is typically $n - 1$ except when calculating population percentiles. When calculating the percentiles (such as the median), typically, there will be one degree of freedom for the estimated standard error because there are usually only two distinct values for the pseudovalues (see homework problem 2). For percentile standard errors, the bootstrap formula is a better approach and is discussed in the next section.

13.5 BOOTSTRAPPING

Bootstrapping is the most modern and perhaps the most widely used method for estimating SE. While there are many variants of the bootstrapping method, there are two main approaches: case bootstrapping and residual bootstrapping (see, e.g., Efron and Tibshirani, 1993). Each has its advantages, but in this textbook we will specifically focus on case bootstrapping. Residual bootstrapping is mainly used to find standard errors in regression models where it is important to keep all the original predictors in the model and there exists a well-accepted, presumed model. Case bootstrapping can be used to find standard errors for almost any statistical procedure where observations are independent. Given that this is a central assumption throughout the book, the case bootstrapping approach can be used for any model described, including models on the univarate, bivariate, or multivariate data and regression. Suppose that one were interested in finding the 85th percentile speed from loop detectors. Furthermore, assume that the measured speeds are independent. In this situation, case bootstrapping would be a good choice.

The algorithm for the case bootstrap will be described below and then demonstrated with an example. Suppose that we have data X_i, $i = 1, \ldots, n$ (e.g., speed measurements), and an estimator $\hat{\theta}(X_1, \ldots, X_n)$ of a transportation performance measure (e.g., the 85th percentile speed). The case bootstrap proceeds as follows:

1. Using the original sample, X_i, $i = 1, \ldots, n$, draw independently with replacement a sample of size n and do this J times. Accordingly, if

the original sample consists of three numbers—2, 5, and 9—the first redrawn sample might be 5, 5, and 2. Denote the redrawn sample as X_{ij}^*, $i=1,\ldots,n$, $j=1,\ldots,J$.

2. For the redrawn sample, X_i^*, $i=1,\ldots,n$, calculate $\hat{\theta}(X_{1j}^*,\ldots,X_{nj}^*)$, $j=1,\ldots J$.

3. Using $\hat{\theta}(X_{1j}^*,\ldots,X_{nj}^*)$, $j=1,\ldots J$, calculate

$$SE(\hat{\theta})=\sqrt{\frac{\sum_{j=1}^{J}(\hat{\theta}(X_{1j}^*,\ldots,X_{nj}^*)-\overline{\theta})^2}{J}}.$$

This is the case resampling bootstrap estimator for the standard error of $\hat{\theta}(X_1,\ldots,X_n)$.

Example 13.3

Suppose that a traffic engineer modeled the relationship between annual hours of delay and average miles traveled on principal arterial streets. She subsequently fitted a linear relationship between the two by using simple linear regression. In this case, the standard formula for calculating standard errors (as presented in Chapter 10) could be used because a linear model was chosen to illustrate the approach. The JMP output is shown in Figure 13.2.

We can see from the JMP output that the standard errors for the intercept and the slope are 151.4 and .19, respectively. In contrast, she may have chosen to use a bootstrap technique to calculate these standard errors. In this case, one hundred case bootstrap samples were drawn for these data, and the bootstrap standard errors for intercept and slope are 157.3 and .22, respectively. The bootstrap estimates were calculated using the MATLAB Statistics Toolbox (2009); this Toolbox has a command for easily calculating bootstrap estimates. Because we are dealing with a simple straight line, the SE estimates from the JMP output in Figure 13.2 are smaller than the bootstrap estimates. Yet, they are also more accurate, as they were from a closed-form solution. However, as is typical, the estimates from both methods are close. As an aside, the estimates provided in the JMP are a trivial case of the propagation of error technique. However, the reader should keep in mind that the bootstrap estimates are typically more accurate than propagation of error for nonlinear estimation of parameters (Efron, 1992). Notice that R^2 equals .76 (from Figure 13.2), but JMP does not provide a standard error for R^2. The bootstrap standard error is .081 for R^2.

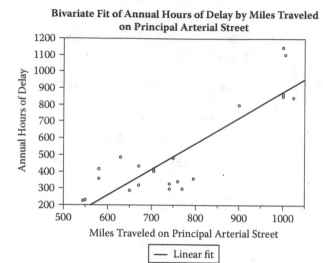

Bivariate Fit of Annual Hours of Delay by Miles Traveled on Principal Arterial Street

Linear Fit

Annual Hours of Delay = −664.6465 + 1.5443783*Miles Traveled on Principal Arterial Street

Summary of Fit

RSquare	0.759012
RSquare Adj	0.746963
Root Mean Square Error	141.8125
Mean of Response	512.5909
Observations (or Sum Wgts)	22

Analysis of Variance

Source	DF	Sum of Squares	Mean Square	F Ratio
Model	1	1266815.6	1266816	62.9918
Error	20	402215.7	20111	Prob > F
C. Total	21	1669031.3		<.0001*

Parameter Estimates

Term	Estimate	Std Error	t Ratio	Prob>|t|
Intercept	-664.6465	151.3777	−4.39	0.0003*
Miles Traveled on Principal Arterial Street	1.5443783	0.194586	7.94	<.0001*

FIGURE 13.2 Partial JMP output for bivariate fit of annual hours of delay by average miles traveled on principal arterial street.

Remark 13.1

The bootstrap method can also be used for calculating confidence intervals using bootstrap percentiles. For example, if one thousand bootstrap samples are drawn for estimating the sample mean, then the confidence interval can be taken as the interval that contains the 26th smallest bootstrap replication of the sample mean to the 975th smallest bootstrap replication of the sample mean. If confidence intervals are obtained in this manner, the bootstrap correctly handles the effects of nonnormal errors. Similar statements apply for prediction intervals.

13.6 CONCLUDING REMARKS

This chapter focused on alternative techniques for estimating a standard error when traditional approaches, which were the focus of much of this textbook, do not apply. These typically occur when (1) the relationship between variables is nonlinear and no reasonable transformation is available to convert the problem to a linear one, or (2) there is no closed-form formula for standard error. It is important to note that these two conditions are often found in transportation problems, and these techniques are useful in situations where bounds on the estimates are required. For example, while it is often constructive to provide predictions, many decision makers would like to know their accuracy. As was shown throughout the textbook, standard errors are instrumental in identifying these bounds: through either confidence or prediction intervals.

The three techniques discussed in this chapter—cross-validation, jack-knifing, and bootstrapping—are highly powerful and relatively easy to use. More importantly, they will allow users to accurately set bounds (i.e., standard errors) on their estimates when standard approaches discussed in previous chapters will not work. This is critically important for many modern transportation applications.

HOMEWORK PROBLEMS

1. Prove that the jackknife calculations for Examples 13.1 and 13.2 are correct.

2. Take a random sample size of ten from a normal distribution using JMP.

   a. Use the jackknife to estimate the standard error for the median. How many degrees of freedom should this standard error have?

   b. Write a computer program to take one hundred bootstrap samples from the ten values that you have and estimate the standard error.

   c. Which estimate of standard error do you prefer?

3. Take a random sample of twenty-five lognormal random variables using JMP. Write a computer program to take one thousand bootstrap samples to calculate the sample mean from each of these samples.

    a. Construct a 95% confidence interval using a bootstrap estimate of standard error.

    b. Construct a 95% confidence interval using a bootstrap estimate of the 2.5th and 97.5th percentiles.

    c. Compare your answers.

4. List three estimators that are not referred to in this book for which you do not have a closed-form formula for standard error.

REFERENCES

Efron, B. 1992. Six questions raised by the bootstrap. In *Exploring the limits of bootstrap*, ed. R. LePage and L. Billard. New York: John Wiley & Sons, pp. 99–125.

Efron, B., and R. J. Tibshirani. 1993. *An introduction to the bootstrap*. London: Chapman & Hall.

Eisele, W. L. 2001. Estimating travel time mean and variance using ITS systems data for real-time and off-line transportation applications. PhD dissertation, Department of Civil Engineering, Texas A&M University, College Station.

Eisele, W. L., and L. R. Rilett. 2001. *Travel time estimates obtained from intelligent transportation systems and instrumented test vehicles: A statistical comparison*, 8–16. Transportation Research Record 1804. Washington, DC: National Research Council.

Green, P. J., and B. W. Silverman. 1994. *Nonparametric regression and generalized linear models*. London: CRC Press.

MATLAB Statistics Toolbox 7.2. 2009. Mathworks, Inc. http://www.mathworks.com/products/statistics/ (accessed December 17, 2009).

Mosteller, F., and J. W. Tukey. 1977. *Data analysis and regression: A second course in statistics*. Reading, MA: Addison-Wesley.

Tukey, J. W. 1958. Bias and confidence intervals in not quite large samples [Abstract]. *Annals of Mathematical Statistics* 29:614.

CHAPTER 14

Bayesian Approaches to Transportation Data Analysis

14.1 INTRODUCTION

In this chapter, we will present the basic concepts and principles of Bayesian approaches as well as simple examples to illustrate the concepts. In previous chapters we discussed exploratory plotting, estimation, and hypothesis testing based upon field studies or designed experiments. Often transportation researchers have more information about mobility, signaling, safety, and materials than just the data at hand. By using a prior distribution, that information can be used in a mathematically rigorous fashion. This is particularly important when sample sizes are small and standard errors are large.

Bayesian estimators are a weighted average among prior opinion and data-driven estimators. They enable researchers to use all the information available, providing a distinct benefit, since well-informed researchers generally estimate better models when they incorporate their knowledge. Researchers that have less experience, however, will benefit by using a noninformative prior. In such cases, a state of vague knowledge can also be used to get Bayesian estimators. Often those estimators will not differ much from the estimators given in the previous chapters.

Finally, we emphasize that this is a quick introduction to Bayesian estimation. References to more extensive coverage of Bayesian statistics and modern advanced procedures are provided. Before researchers can appreciate more advanced approaches, it is useful to have a good grasp of the material presented in this chapter.

14.2 FUNDAMENTALS OF BAYESIAN STATISTICS

In Bayesian approaches the parameters are viewed as random variables having their own distributions, as opposed to classical (or frequentist) approaches treating the unknown parameters as fixed constants. The distribution of parameters assumed before observing any data is called a prior distribution. The key idea of a Bayesian approach is that all (prior) knowledge or extra information in addition to the data about the parameters can be incorporated into the estimation through the prior distributions on model parameters. Let θ be a parameter of interest. For example, it could be a true free-flow mean speed along a highway corridor. Suppose that we have information from a previous study that the mean speed along this corridor is approximately 55 mph with a standard deviation of approximately 3 mph. The prior distribution on θ may be assumed to be normal with a mean of 55 and standard deviation of 3, i.e., $\theta \sim N(55, 3^2)$.

The prior information on the parameter can be updated by the information contained in the data through Bayes' theorem. In Chapter 4, we introduced Bayes' theorem in terms of discrete events/probabilities. It can also, however, be stated in terms of continuous probability distributions. The two main uses for Bayes' theorem (also known as Bayes' rule) are (1) to calculate conditional probabilities that would be difficult to do correctly, and (2) to provide improved parameter estimates that combine various sources of information. Discussion on the second use has been deferred until the present chapter.

In Bayesian statistics, Bayes' theorem plays a major role. Once the data are obtained, the information contained in the data can be combined with the prior distribution through Bayes' theorem. This leads to an updated distribution for the parameter, called a posterior distribution. Let us denote the sampling distribution (or the likelihood, see the appendix of Chapter 11) by $f(x|\theta)$, the prior distribution by $p(\theta)$, and the posterior distribution by $\pi(\theta|x)$. Bayes' theorem states that

$$\pi(\theta|x) = \frac{f(x|\theta)p(\theta)}{m(x)} \tag{14.1}$$

where $m(x)$ is the marginal distribution of the data (x) given by

$$m(x) = \int f(x|\theta)p(\theta)d\theta. \qquad (14.2)$$

The marginal distribution $m(x)$ is actually the normalizing constant that makes the integral of $\pi(\theta|x)$ equal to 1. Because $m(x)$ does not depend on θ, Bayes' theorem is sometimes stated more briefly as

$$\pi(\theta|x) \propto f(x|\theta)p(\theta) \qquad (14.3)$$

or

$$posterior \propto likelihood \times prior.$$

The $\propto$ means "is proportional to." Note that the posterior distribution (Equation 14.1) is the conditional distribution of θ given the data. Bayesian inferences (which will be discussed in the next section) are made based on the posterior distribution. The ability to incorporate various sources of information (the information contained in the data and prior information) into parameter estimation leads to an improved estimate. This is a strength of a Bayesian approach.

The use of Bayesian statistics in transportation has recently become more popular. See, for example, Park et al. (2008b) for a Bayesian approach for pavement performance prediction and Schluter et al. (1997), Carriquiry and Pawlovich (2004), Park et al. (2010), and references therein for fully Bayesian approaches in traffic safety analysis. Ang and Tang (2006) presented a good overview on the use of Bayesian statistics in transportation. Readers who desire more extensive coverage and discussions of Bayesian statistics in the general context are referred to Bolstad (2007), DeGroot and Schervish (2002), and Lee (2004). Part of the challenge with the use of Bayesian statistics is a computational difficulty. Except for some simple models with special types of priors called conjugate priors (which will be introduced in the next section), Bayesian computation can typically not be done analytically, and oftentimes numerical approximation or simulation is required. Computational techniques that have become very popular in modern Bayesian statistics are Markov chain Monte Carlo (MCMC) methods, which is beyond the scope of this textbook. The MCMC approach allows for the simulation of complex, nonstandard multivariate distributions by generating random variables from a (marginal) distribution indirectly, which obviates the need to calculate the density. After a sufficiently large number of iterations, samples generated from MCMC can be regarded as posterior samples (i.e., samples generated from the posterior distribution). Two of the widely used MCMC algorithms are Gibbs sampling and the Metropolis–Hastings algorithm. For a general overview and various usages of MCMC (in nontransportation contexts), interested readers are referred to, for example, Gilks et al. (1996) and Robert and Casella (2004). Park et al. (2008a) provide a helpful discussion of the use of MCMC in origin-destination matrix estimation and imputation of missing volumes.

14.3 BAYESIAN INFERENCE

Bayesian inference about the parameter is based on the posterior distribution, which is proportional to the likelihood (distribution of the data) and the prior distribution (distribution of the parameter). Thus, the very first step in Bayesian analysis is to specify the likelihood and the prior distribution. Unlike the likelihood, we can choose relatively easily from several available distributions that best approximate the data; the decision on what prior distribution is best to use is not straightforward. Bolstad (2007) provides helpful guidelines on how to choose a prior in a general statistics context. If there is actually very good prior information (e.g., from the previous study) or expert knowledge on the parameter of interest (e.g., on the range of plausible values), it is best to incorporate such information into the prior distribution. For example, if we know from the previous study that the free-flow average speed (μ) is around 62 mph with a standard deviation of about 3 mph, we may set our prior distribution on μ to be the normal distribution with prior mean 62 and prior standard deviation 3. (Free-flow speeds are known to be well approximated by normal distribution if there are not many outliers.)

Oftentimes, however, there is no precise knowledge on the parameter of interest, because often the purpose of the study is to estimate the unknown parameter values. In these situations vague priors—sometimes referred to as noninformative priors or reference priors—are typically used. A vague prior has a relatively flat distribution compared to the likelihood. For example, a normal distribution with a very large standard deviation such as 100 (when the mean is 62, as above) can be viewed as a noninformative prior because the distribution is almost flat. By *flat* we mean that the distribution assigns approximately equal weight on all possible values of the parameter, which in this case is average speed. In this case, we hope that the effect of priors on the posterior distribution is minimal, because we do not have a very good prior knowledge on the parameter, and the posterior distribution will be largely determined by the likelihood. We should note, however, that the determination of noninformative priors is not always a clear-cut decision, especially in high-dimensional parameter space. In fact, the selection of priors often requires careful thought, particularly when there is a paucity of good prior information. For a more in-depth discussion of this topic, interested readers are referred to Kass and Wasserman (1996). Schluter et al. (1997) also provide a good discussion on the elicitation of both informative and noninformative priors, in crash data analysis.

14.3.1 Conjugate Priors

Recall from Bayes' rule (Equation 14.3) that

$$posterior \propto likelihood \times prior.$$

It would be mathematically convenient if *likelihood* × *prior* leads to a formula for a known distribution from which the posterior distribution can be identified. An easy way to achieve this is to select a prior so that *likelihood* × *prior* is in a similar functional form as the prior.

14.3.1.1 For Poisson Likelihood

Let X be the number of pedestrian arrivals per minute at a crossing with an unknown arrival rate of λ. In Chapter 5 we learned that the Poisson distribution approximates such data well. Thus, we can choose the distribution for X (likelihood) to be Poisson (λ). That is,

$$f(x|\lambda) = \frac{e^{-\lambda}\lambda^x}{x!}.$$

Now we need to choose the prior distribution for λ. Note that λ can be continuous and can only take nonnegative values (because it is an arrival rate). Among the set of continuous distributions having nonnegative supports, the gamma distribution is mathematically tractable and makes *likelihood* × *prior* essentially the same formula as the gamma prior. Recall that the gamma distribution with parameters α and β, $Gamma(\alpha,\beta)$, is given as

$$p(\lambda) = \frac{\lambda^{\alpha-1}e^{-\lambda/\beta}}{\Gamma(\alpha)\beta^{\alpha}}, \lambda \geq 0, \alpha, \beta > 0.$$

By entering the formula for the Poisson likelihood and the gamma prior into Equation 14.3, we get

$$\pi(\theta|x) \propto f(x|\theta)p(\theta)$$

$$\propto (e^{-\lambda}\lambda^{x})(\lambda^{\alpha-1}e^{-\lambda/\beta})$$

$$= \lambda^{x+\alpha-1}e^{-\lambda(1+1/\beta)}$$

which is essentially the same formula as the gamma prior except that the parameters are different (i.e., updated to $x + \alpha$ and $(1 + 1/\beta)^{-1}$). The full gamma posterior density is given as

$$\pi(\lambda|x) = \frac{\lambda^{x+\alpha-1}e^{-\lambda(1+1/\beta)}(1+1/\beta)^{x+\alpha}}{\Gamma(x+\alpha)}. \tag{14.4}$$

Note that a gamma prior gives rise to a posterior distribution that again belongs to a gamma family.

A prior is said to be a conjugate prior if the prior and the posterior distributions belong to the same family, in other words, if the posterior has the same distributional form as the prior distribution. Another commonly used conjugate prior includes a beta prior for the binomial likelihood and a normal prior for the normal likelihood.

14.3.1.2 For Binomial Likelihood

Let n be the number of vehicles entering a particular intersection in a given time and X_i ($i = 1,\ldots,n$) be a random variable taking a value of 1 if the ith vehicle runs the red light and 0 if it does not. Then $X_1,\ldots,X_n$ can be regarded as a sample of independent observations following $Bernoulli(p)$, where p is the unknown true rate of red light running at the intersection. Let Y be the number of vehicles that run red lights out of n

vehicles, i.e., $Y = \sum_{i=1}^{n} X_i$. Then Y follows a binomial distribution, $Bin(n,p)$, with the following likelihood:

$$f(y|p) = \binom{n}{y} p^y (1-p)^{n-y}.$$

Assume that the prior distribution on p is $Beta(\alpha,\beta)$ with the prior density function given as

$$p(p) = \frac{\Gamma(\alpha+\beta)}{\Gamma(\alpha)\Gamma(\beta)} p^{\alpha-1} (1-p)^{\beta-1}.$$

It can be shown that the posterior distribution of p is $Beta(y + \alpha, n-y + \beta)$ with the following density function:

$$\pi(p|y) = \frac{\Gamma(n+\alpha+\beta)}{\Gamma(y+\alpha)\Gamma(n-y+\beta)} p^{y+\alpha-1}(1-p)^{n-y+\beta-1} \qquad (14.5)$$

where $0 \le p \le 1$, which belongs to the same family in distribution (beta distribution) as the prior (see homework problems for details). Note that we merely need to add the number of successes y to α and the number of failures $n - y$ to β to obtain the parameters of the posterior distribution in this case. From the property of a beta distribution, it can be shown that the mean and variance of the posterior distribution in Equation 14.5 are

$$E(p|y) = \frac{y+\alpha}{\alpha+\beta+n} \qquad (14.6)$$

and

$$var(p|y) = \frac{(y+\alpha)(n-y+\beta)}{(n+\alpha+\beta)^2 (n+\alpha+\beta+1)}. \qquad (14.7)$$

14.3.1.3 For Normal Likelihood

For the normal likelihood, a conjugate prior is also normal, which will lead to a normal posterior distribution. To illustrate, let $X_1,...,X_n$ be a sample of independent observations distributed $N(\mu,\sigma^2)$ where σ^2 is known. Let us assume that the prior distribution on μ is $N(m,\tau^2)$, where m and τ^2 are known. Then it can be shown that the posterior distribution of μ is normal with mean and variance given by

$$E(\mu|\bar{x}) = \left(\frac{n}{\sigma^2}\bar{x} + \frac{m}{\tau^2}\right) \Big/ \left(\frac{n}{\sigma^2} + \frac{1}{\tau^2}\right) = \frac{\tau^2}{\frac{\sigma^2}{n} + \tau^2}\bar{x} + \frac{\frac{\sigma^2}{n}}{\frac{\sigma^2}{n} + \tau^2}m \quad (14.8)$$

and

$$Var(\mu|\bar{x}) = \left(\frac{n}{\sigma^2} + \frac{1}{\tau^2}\right)^{-1} = \frac{\frac{\sigma^2}{n}\tau^2}{\frac{\sigma^2}{n} + \tau^2}. \quad (14.9)$$

See Casella and Berger (1990) for details.

It can be seen from Equation 14.8 that the posterior mean of a normal population mean, μ, can be viewed as a weighted average of the sample mean and the prior mean with weights inversely proportional to the variance of $\bar{X}$ and the prior variance, respectively.

Example 14.1

Let X_i ($i = 1,...,n$) denote the number of fatal crashes per year at the ith intersection. Say that $X_1,...,X_n$ are independent and each follows a Poisson distribution with the mean crash rate λ. Assume that the mean crash rate is not constant, but is believed to be a gamma-distributed random variable possessing a prior density with parameters α and β. That is, $\lambda \sim Gamma(\alpha,\beta)$. What would be the posterior distribution after observing $x_1,...,x_n$ crashes from n intersections?

By Bayes' rule,

$$\pi(\lambda|x_1,\cdots,x_n) \propto f(x_1,\cdots,x_n|\lambda)p(\lambda)$$

where

$$f(x_1,\cdots,x_n|\lambda) = \prod_{i=1}^{n} f(x_i|\lambda) = \prod_{i=1}^{n} \frac{\lambda^{x_i}e^{-\lambda}}{x_i!} = \frac{\lambda^{\sum_{i=1}^{n}x_i}e^{-n\lambda}}{x_i!}$$

and

$$p(\lambda) = \frac{1}{\Gamma(\alpha)\beta^\alpha}\lambda^{\alpha-1}e^{-\lambda/\beta}.$$

By retaining the terms that involve λ only, it can be shown that

$$\pi(\lambda|x_1,\cdots,x_n) \propto \lambda^{\sum_{i=1}^{n} x_i} e^{-n\lambda} \lambda^{\alpha-1} e^{-\lambda/\beta}$$

$$= \lambda^{\sum_{i=1}^{n} x_i+\alpha-1} e^{-\lambda(n+1/\beta)}$$

which looks like a gamma distribution with parameters $\sum_{i=1}^{n} x_i+\alpha$ and $(n+1/\beta)^{-1}$ except for a normalizing constant. The full posterior density of λ with a normalizing constant is

$$\pi(\lambda|x_1,\cdots,x_n) = \frac{\lambda^{\sum_{i=1}^{n} x_i+\alpha-1} e^{-\lambda(n+1/\beta)} (n+1/\beta)^{\sum_{i=1}^{n} x_i+\alpha}}{\Gamma\left(\sum_{i=1}^{n} x_i + \alpha\right)} \qquad (14.10)$$

which can be denoted by $Gamma\left(\sum_{i=1}^{n} x_i+\alpha,(n+1/\beta)^{-1}\right)$. The reader should note that the prior distribution on the mean crash rate λ, $Gamma(\alpha,\beta)$, has been updated to the posterior distribution, $Gamma\left(\sum_{i=1}^{n} x_i+\alpha,(n+1/\beta)^{-1}\right)$, after observing the data. The mean and variance of the posterior distribution in Equation 14.10 are given as

$$E(\mu|x_1,\cdots,x_n) = \left(\sum_{i=1}^{n} x_i+\alpha\right)(n+1/\beta)^{-1} \qquad (14.11)$$

and

$$E(\mu|x_1,\cdots,x_n) = \left(\sum_{i=1}^{n} x_i+\alpha\right)(n+1/\beta)^{-2} \qquad (14.12)$$

respectively.

Table 14.1 contains the list of commonly used conjugate priors for some of the distributions that are often used in transportation studies.

We should note that one of the main motivations for using a conjugate prior is computational convenience. By using a conjugate prior, it is possible to obtain a closed-form formula for a posterior distribution and a Bayesian estimator (that will be presented in the next section). There are cases, however, where a conjugate prior may not exist for a given model or is not appropriate to summarize the scientific knowledge available *a priori* for the parameter of interest. In these situations other computational techniques, such as MCMC, which was mentioned earlier in this chapter, can be employed to obtain the samples from the posterior distribution.

TABLE 14.1 Examples of Conjugate Priors

Likelihood	Parameter of Interest	Conjugate Prior	Posterior Distribution
Binomial	p	Beta(α, β)	Beta$\left(\sum_{i=1}^{n} x_i + \alpha, \, n - \sum_{i=1}^{n} x_i + \beta\right)$
Poisson	λ	Gamma(α, β)	Gamma$\left(\sum_{i=1}^{n} x_i + \alpha, \, (n + 1/\beta)^{-1}\right)$
Negative binomial	p	Beta(α, β)	Beta$\left(\alpha + rn, \, \sum_{i=1}^{n} x_i + \beta\right)$
Geometric	p	Beta(α, β)	Beta$\left(\alpha + n, \, \sum_{i=1}^{n} x_i + \beta\right)$
Normal with known variance σ^2	μ	Normal(m, τ^2)	Normal$\left(\left(\dfrac{n\bar{x}}{\sigma^2} + \dfrac{m}{\tau^2}\right) \Big/ \left(\dfrac{n}{\sigma^2} + \dfrac{1}{\tau^2}\right), \left(\dfrac{n}{\sigma^2} + \dfrac{1}{\tau^2}\right)^{-1}\right)$
Normal with known mean μ	σ^2	IG(α,β)	$IG\left(\alpha + \dfrac{n}{2}, \left[\dfrac{1}{\beta} + \dfrac{\sum_{i=1}^{n}(x_i-\mu)^2}{2}\right]^{-1}\right)$
Normal with unknown mean μ	σ^2	IG(α,β)	$IG\left(\alpha + \dfrac{n-1}{2}, \left[\dfrac{1}{\beta} + \dfrac{(n-1)S^2}{2}\right]^{-1}\right)$

Note: (1) $IG(\alpha,\beta)$ denotes the inverse gamma distribution with parameters α and β. (2) σ^2 is said to follow $IG(\alpha,\beta)$ when $1/\sigma^2$ follows $Gamma(\alpha,\beta)$. (3) S^2 is the sample variance based on a sample of size n.

This means that a closed-form formula for the posterior distribution does not need to be obtained. The posterior summaries are computed based on those samples. Interested readers are referred to the references given in the previous section.

14.3.2 Point Estimation

A Bayesian approach may utilize various summaries of the posterior distribution for estimation of the parameter. The mean of the posterior distribution (posterior mean) is often used as the Bayes estimator of the parameter. The definition of the posterior mean is given by

$$E(\theta|x) = \int \theta \pi(\theta|x) d\theta. \qquad (14.13)$$

For the conjugate distributions of Table 14.1, the posterior mean can be analytically obtained and a closed-form solution is known. Other summaries of the posterior distribution, such as the posterior median, posterior mode, or posterior percentiles, may also be used as a Bayesian point estimator.

Example 14.2

Researchers are interested in estimating the true free-flow mean speed (μ) along a corridor. Say that from the previous studies the mean speed can be assumed to follow a normal distribution with a mean of 55 and standard deviation of 3, i.e., $\mu \sim N(55,3^2)$. Suppose that researchers observed five speeds ranging from 39 to 53 from $X \sim N(\mu,6^2)$, and the sample mean speed value was determined to be 45, which was significantly lower than the prior mean speed (55). The prior mean (55) can be updated by incorporating the information obtained from the data (sample mean speed of 45) to the posterior mean (Bayes estimate). Recall that the posterior mean is given by a weighted average of the maximum likelihood estimate (sample mean) and the prior mean. Using the formula given in Equation 14.8, the Bayes estimate can be calculated as

$$E(\mu|\bar{x}) = \frac{\tau^2}{\frac{\sigma^2}{n}+\tau^2}\bar{x} + \frac{\frac{\sigma^2}{n}}{\frac{\sigma^2}{n}+\tau^2}m = \frac{3^2}{\frac{6^2}{5}+3^2} \times 45 + \frac{\frac{6^2}{5}}{\frac{6^2}{5}+3^2} \times 55 = 49.44.$$

The Bayes estimate of the mean speed (49.44) lies between the sample mean (45) and the prior mean (55). Compared to the maximum likelihood estimate (MLE), the Bayes estimate is an improved estimate in that it incorporates external information (prior information). Note that, in this case, the maximum likelihood estimate (45) and the Bayes estimate (49.44) differ considerably because

n is small and the sample values differ markedly from those that would be expected given the prior distribution. As the sample size increases, however, the difference between two estimators decreases in general.

Example 14.3

Let n be the number of vehicles entering a particular intersection in a given time, Y the number of vehicles that run red lights out of n vehicles, and p the unknown true violation rate of red light running at the intersection. Researchers are interested in estimating the violation rate at the intersection. Based on the historical data, the violation rate is assumed to follow a beta distribution with the parameters $\alpha = 2$ and $\beta = 7$. Then the prior mean violation rate (the probability of red light running anticipated before the observation of any data) is

$$E(p) = \frac{\alpha}{\alpha + \beta} = 0.2222.$$

Say that twenty vehicles entered the intersection during the five-minute time interval and two of the vehicles ran a red light. How can this observation be used to update the prior distribution of probability of red light running, *Beta* (2, 7)? What is the Bayes estimate of p? How different is it from the expected probability of red light running before observing any data (0.2222)?

First note that Y (the number of vehicles that run red lights out of n vehicles) follows a binomial distribution, $Y \sim Bin(n,p)$. Recall that the posterior distribution of p with the Beta prior is $Beta(y + \alpha, n-y + \beta)$. In our case $n = 20$, $y = 2$, $\alpha = 2$, and $\beta = 7$. Entering those numbers in $Beta(y + \alpha, n-y + \beta)$, we get the posterior distribution of p as $Beta(4,25)$. From Equation 14.6, the posterior mean (Bayes estimate) of p can be shown to be

$$E(p|y) = \frac{y + \alpha}{n + \alpha + \beta} = \frac{2+2}{20+2+7} = 0.1379.$$

The Bayes estimate of p (0.1379) lies between the sample mean of the underlying Bernoulli trials, $y/n = 0.1$, and the prior mean of p, $E(p) = \alpha/(\alpha + \beta) = 0.2222$. Note that, after observing the data, the information in the data has been incorporated into the prior mean violation rate and led to an updated estimate (posterior mean violation rate) that combines both sources of information.

The posterior mean of p can be expressed as a weighted average of the sample mean (y/n) and the prior mean $\alpha/(\alpha + \beta)$ with weights depending on α, β, and n as follows:

$$E(p|y) = \frac{y + \alpha}{n + \alpha + \beta} = \left(\frac{n}{n + \alpha + \beta}\right)\left(\frac{y}{n}\right) + \left(\frac{\alpha + \beta}{n + \alpha + \beta}\right)\left(\frac{\alpha}{\alpha + \beta}\right). \tag{14.14}$$

For the Bayes estimate (0.1379) of Example 14.3, we have

$$0.1379 = \left(\frac{20}{20+2+7}\right)\left(\frac{2}{20}\right) + \left(\frac{2+7}{20+2+7}\right)\left(\frac{7}{2+7}\right)$$

$$= (0.6897)(0.1) + (0.2414)(0.2222).$$

That is, the Bayes estimate for the probability of running a red light (0.1379) is the weighted average of the sample proportion of vehicles running a red light (0.1) and the prior estimate of the probability (0.2222), where the weights are given as 0.6897 and 0.2414 in this case.

It can be seen from Equation 14.14 that as the sample size (n) increases, the MLE will get proportionately more weight, and as a result, the Bayes estimate (posterior mean) would become closer to the MLE (sample mean). On the other hand, when the sample size is small, the prior mean is given more weight and the Bayes estimate would be closer to the prior mean. This is when the role of the prior becomes more important. Although the use of good prior knowledge will improve the Bayes estimate and will lead to a more precise estimate regardless of the sample size, the effect will become more prominent when the sample size is small. At the same time, it needs to be remembered that the use of wrong or incorrect prior information when the sample size is small would affect the resulting Bayes estimate more significantly than when the sample size is large. While the advantage of Bayesian approaches becomes clearer in the small sample case, that is when the use of good prior information and selection of reasonable priors becomes more crucial, so as not to generate misleading results.

14.3.3 Uncertainty Estimation

As in the classical analysis, it is important to know not only the point estimate for the parameter but also how good the estimate is with respect to precision and accuracy. Put another way, the point estimate itself may not be very useful without knowing how much uncertainty is associated with it. In Bayesian analysis, a measure of variability of the posterior distribution can be used as an uncertainty estimate for the parameter (or an estimate for the uncertainty of the parameter estimate). An interval covering the central area of the posterior distribution can also be used as an uncertainty estimate.

14.3.3.1 Posterior Standard Deviation

The standard deviation of the posterior distribution (posterior standard deviation) describes the uncertainty in the parameter in Bayesian inference. The posterior standard deviation can be thought of as a counterpart of the standard error used in the classical approaches. The definition of the posterior standard deviation is given by

$$Std(\theta|x) = [Var(\theta|x)]^{1/2}$$

$$= \left[\int (\theta - E(\theta|x))^2 \pi(\theta|x) d\theta \right]^{1/2}. \tag{14.15}$$

Other summaries of the posterior distribution, such as the posterior inter-quartile range or posterior variance, may also be used to provide uncertainty estimates for θ.

Example 14.4 (Example 14.2 continued)

Researchers are also interested in knowing how precise their estimate for the free-flow mean speed (49.44) is. The uncertainty estimate for the Bayes estimate of the free-flow mean speed μ can be obtained as follows. From Equation 14.9, the posterior variance is

$$Var(\mu|\bar{x}) = \left(\frac{n}{\sigma^2} + \frac{1}{\tau^2} \right)^{-1} = \frac{\frac{\sigma^2}{n} \tau^2}{\frac{\sigma^2}{n} + \tau^2} = \frac{\frac{6^2}{5} \times 3^2}{\frac{6^2}{5} + 3^2} = 4.$$

Then the posterior standard deviation is

$$Std(\mu|\bar{x}) = \sqrt{Var(\mu|\bar{x})} = 2.$$

Example 14.5 (Example 14.3 continued)

The Bayes estimate for the probability of running a red light (0.1379) is subject to uncertainty. The uncertainty estimate for the Bayes estimate of p can be obtained by the standard deviation of the posterior distribution as follows. From Equation 14.7, the posterior variance of $Beta(y + \alpha, n-y + \beta)$ is

$$Var(p|y) = \frac{(y+\alpha)(n-y+\beta)}{(n+\alpha+\beta)^2(n+\alpha+\beta+1)} = \frac{(1+2)(20-1+7)}{(20+2+7)^2(20+2+7+1)} = 0.0031.$$

Thus, the posterior standard deviation is

$$Std(p|y) = \sqrt{Var(p|y)} = \sqrt{0.0031} = 0.0556.$$

14.3.3.2 Credible Intervals

Credible intervals used in Bayesian interval estimation may be viewed as a counterpart of confidence intervals used in classical interval estimation, but with a very different interpretation. Given a posterior distribution $\pi(\theta|x)$, C is a credible interval for θ if

$$P(\theta \in C|x) = \int_C \pi(\theta|x)d\theta. \tag{14.16}$$

For example, a 95% credible interval for θ can be constructed by finding an interval C over which $\int_C \pi(\theta|x)d\theta = 0.95$. A probability interpretation such that there is 95% probability that θ is inside the interval C can be attached to Bayesian credible intervals unlike the confidence intervals. This is in contrast to confidence intervals where the confidence level associated with the confidence interval concerns the long-run percentages of similarly constructed intervals covering the true parameter value from similar independent experiments. There are many different ways to form a credible interval (see Casella and Berger, 1990). Credible intervals with equal tails are frequently used in Bayesian interval estimation. A $(1-\alpha)\times100\%$ equal-tail credible interval corresponds to the $100(\alpha/2)$th and $100(1-\alpha/2)$th percentiles of the posterior distribution. For example, the 2.5th percentile and the 97.5th percentile can be used to construct the 95% credible interval.

For normal mean μ (with known variance σ^2), a $(1-\alpha)\times100\%$ Bayesian credible interval (obtained by using a normal prior) can be given as

$$E(\mu|\bar{x}) \pm z_{1-\alpha/2} Std(\mu|\bar{x}) \tag{14.17}$$

where $E(\mu|\bar{x})$ and $Std(\mu|\bar{x})$ are posterior mean and posterior standard deviation (see Equations 14.8 and 14.9), respectively, and $z_{1-\alpha/2}$ is the upper $\alpha/2$ percentile of the standard normal distribution.

Remark 14.1

If σ^2 is unknown, the sample variance s^2 can be used in place of σ^2 in Equations 14.8 and 14.9 to obtain $E(\mu|\bar{x})$ and $Std(\mu|\bar{x})$. The credible interval given in Equation 14.17 then should be widened to account for added uncertainty due to replacing σ^2 by the sample variance s^2. The $(1-\alpha)\times100\%$ Bayesian credible interval for μ in that case is given as

$$E(\mu|\bar{x}) \pm t_{n-1,1-\alpha/2} Std(\mu|\bar{x}) \tag{14.18}$$

where $t_{n-1,1-\alpha/2}$ is the upper $\alpha/2$ percentile of the t-distribution with $n-1$ degrees of freedom.

Example 14.6 (Example 14.4 continued)

Researchers desire to find an interval that is believed to contain the free-flow mean speed at some probability level. Say that they wish to obtain a 95% credible interval for μ. Because the standard deviation of speeds is known to be $\sigma = 6$ in this case, the formula in Equation 14.17 can be used to obtain the credible interval. Recall that the posterior mean and posterior standard deviation were $E(\mu|\bar{x}) = 49.44$ and $Std(\mu|\bar{x}) = 2$, respectively, from Examples 14.2 and 14.4. By plugging in those values as well as $z_{1-\alpha/2} = 1.96$ in Equation 14.17, their 95% credible interval can be obtained as

$$E(\mu|\bar{x}) \pm z_{1-\alpha/2} Std(\mu|\bar{x}) = 49.44 \pm (1.96)(2) = 49.44 \pm 3.92.$$

The lower limit and upper limit of the interval are 45.52 and 53.36. The 95% credible interval [45.52, 53.36] is an interval that has a posterior probability of 95% of containing μ. That is, there is a 95% probability that μ belongs to [45.52, 53.36].

Example 14.7 (Example 14.3 continued)

We can also find a 95% credible interval for the probability of running red light p. Recall from Example 14.3 that the posterior distribution of p was $Beta(4,25)$. To get the lower and upper limit of the 95% credible interval, we only need to find the 2.5th and 97.5th percentiles from $Beta(4,25)$ distribution. Those values can easily be computed by JMP (see Section 5.3.1) or any other statistical software package. The 2.5th and 97.5th percentiles from $Beta(4,25)$ are 0.040 and 0.282, respectively. The 95% credible interval for p is [0.040, 0.282], which has a posterior probability of 95% of containing p.

Example 14.8

Suppose that in Example 14.3 researchers did not have any prior information or expectation on the probability of red light running, p, before they started their study. Thus, they decided to use the uniform prior that gives equal weight to all possible values of p in [0, 1]:

$$p(p) = 1 \text{ for } 0 \leq p \leq 1. \tag{14.19}$$

The prior in Equation 14.19 can be viewed as a noninformative prior for p, as it does not favor any one value over another. By Bayes' theorem, the posterior distribution for p in this case is

$$\pi(p|y) = \binom{n}{y} p^y (1-p)^{n-y} \times 1 \tag{14.20}$$

for $0 \leq p \leq 1$. Equation 14.20 can be rewritten as

$$\pi(p|y) = \frac{\Gamma(n+1+1)}{\Gamma(y+1)\Gamma(n-y+1)} p^{y+1-1}(1-p)^{n-y+1-1}$$ (14.21)

which corresponds to the density of the beta distribution $Beta(y+1,n-y+1)$. In fact, the uniform prior in Equation 14.19 is a special case of the $Beta(\alpha,\beta)$ prior where $\alpha = 1$ and $\beta = 1$. The posterior distribution of p with $n = 20$ and $y = 2$ is $Beta(3,19)$. The posterior mean and variance of p can be obtained by substituting in $n = 20$, $y = 2$, $\alpha = 1$, and $\beta = 1$ in the formulas of Equations 14.6 and 14.7 as follows:

$$E(p|y) = \frac{y+\alpha}{\alpha+\beta+n} = \frac{2+1}{1+1+20} = \frac{3}{22} = 0.1364$$

and

$$var(p|y) = \frac{(y+\alpha)(n-y+\beta)}{(n+\alpha+\beta)^2(n+\alpha+\beta+1)} = \frac{(2+1)(20-2+1)}{(20+1+1)^2(20+1+1+1)} = 0.0051.$$

The posterior standard deviation is

$$Std(p|y) = \sqrt{Var(p|y)} = \sqrt{0.0051} = 0.0716.$$

We can see that the Bayes estimate of p in this case is related to the frequentist estimate of p obtained by adding one success and one failure in Chapter 9.

Recall from Examples 14.3 and 14.5 that the posterior mean and standard deviation of p were 0.1379 and 0.0556, respectively, when the prior was $Beta(2,7)$. Note that the Bayes estimate obtained by using a noninformative prior (0.1364) is still relatively close to the Bayes estimate obtained by using $Beta(2,7)$ prior (0.1379). However, the posterior standard deviation under the noninformative prior (0.0716) is larger than that under the $Beta(2,7)$ prior (0.0556). The larger uncertainty estimate under the noninformative prior is a natural consequence of not incorporating any prior information on p into parameter estimation. The 95% credible interval for p under the noninformative prior can be obtained as the 2.5th and 97.5th percentiles from $Beta(3,19)$. The resulting interval [0.030, 0.304] is, as expected, wider than the 95% credible interval of Example 14.7.

14.4 CONCLUDING REMARKS

Often transportation professionals have very good information on a process before they begin a study. For example, they may have some information on the physical properties of a material based on published material, or they may have information on the demand for transportation services based on earlier studies. Until now, this textbook concentrated on approaches that assumed that this information did not exist—or at least would not be useful for estimating models. However, the approaches discussed explicitly use this information (e.g., the prior information) with the newly observed data to develop statistical models. These techniques are particularly useful when (1) the prior information is good or trustworthy, and (2) the amount of newly collected data is limited. Because they have been shown to be very useful in other fields (e.g., not wasting information), these Bayesian approaches have become increasingly popular in transportation studies.

In this chapter we showed how researchers can combine prior knowledge with newly collected data to develop Bayesian models. This in turn can be used to estimate transportation phenomena of interest, such as the proportion of drivers who run red lights at signalized intersections. We showed, by example, that these estimates are easily computed with the use of conjugate priors, and follow understandable mathematical principles. We presented Bayesian credible intervals based upon posterior distributions. Bayesian estimators are widely used in safety analyses, although the models published in the literature mostly use nonconjugate priors because of complexity issues. Although the general theory remains the same, closed-form estimators are not (generally) available because prior distributions are not conjugate and posterior distributions are not given in any known form. In such cases, computer simulations must be used to sample from the posterior distribution. A common simulation technique is MCMC. By understanding the process of using conjugate priors, the reader will have a good basis for reading references that use MCMC methods. In some sense this parallels the use of the bootstrap in Chapter 13 to get standard errors when no closed form exists. Thus, this chapter provides readers with an introduction to the concept of Bayes estimation and credible intervals.

HOMEWORK PROBLEMS

1. Let Y be the number of vehicles that run red lights out of n vehicles, following a binomial distribution, $Bin(n,p)$, and let p have a $Beta(\alpha,\beta)$ distribution, the conjugate family for the binomial.

a. Using Bayes' rule, show that the posterior distribution of p, $\pi(p|y)$, is proportional to

$$\binom{n}{y}\frac{\Gamma(\alpha+\beta)}{\Gamma(\alpha)\Gamma(\beta)}p^{y+\alpha-1}(1-p)^{n-y+\beta-1}$$

b. The beta function is defined as $B(\alpha,\beta)=\int_0^1 p^{\alpha-1}(1-p)^{\beta-1}dp$, and the following relationship holds between the beta function and the gamma function:

$$B(\alpha,\beta)=\frac{\Gamma(\alpha)\Gamma(\beta)}{\Gamma(\alpha+\beta)}.$$

Using this relationship, show that

$$\pi(p|y)=\frac{\Gamma(n+\alpha+\beta)}{\Gamma(y+\alpha)\Gamma(n-y+\beta)}p^{y+\alpha-1}(1-p)^{n-y+\beta-1}$$

which is $Beta(y+\alpha,n-y+\beta)$. Researchers are interested in estimating the annual average fatal crashes at twenty intersections in a city. Suppose that the annual fatal crash count follows a Poisson distribution with the mean λ. It is believed that the mean crash count, λ, is not a constant, but a random variable distributed as $Gamma(2,3)$. Researchers collected the data on the fatal crash counts for a year from twenty intersections, and the total fatal crash count (the sum of fatal crash counts over twenty intersections) turned out to be twenty-five.

c. Find the posterior distribution of λ. (Hint: Note that a gamma distribution is the conjugate family for the Poisson. Also, note that $\sum_{i=1}^n x_i = 25$ in Table 14.1. Enter another number in the formula too.)

d. What is the prior mean fatal crash count?

e. Find the Bayes estimate of λ. (Hint: The Bayes estimate is the posterior mean fatal crash count.) How different is it from the prior mean fatal crash count?

f. Find an uncertainty estimate for the Bayes estimate of λ. (Hint: The posterior standard deviation can be used as an uncertainty estimate.)

g. Compute the 95% credible interval for λ. (Hint: You can use JMP or other statistical software packages to get the gamma percentiles.) How would you interpret this interval?

REFERENCES

Ang, A. H. -S., and W. H. Tang. 2006. *Probability concepts in engineering: Emphasis on applications to civil and environmental engineering*. New York, NY: John Wiley & Sons.

Bolstad, W. M. 2007. *Introduction to Bayesian statistics*. 2nd ed. New York: John Wiley & Sons.

Carriquiry, A., and M. D. Pawlovich. 2004. From empirical Bayes to full Bayes: Methods for analyzing traffic safety data. http://www.iowadot.gov/crashanalysis/pdfs/eb_fb_comparison_whitepaper_october2004.pdf (accessed May 7, 2009).

Casella, G., and R. L. Berger. 1990. *Statisfical inference*. Pacific, Grove, CA: Wadsworth Brooks/Cole.

DeGroot, M. H., and M. J. Schervish. 2002. *Probability and statistics*. Reading, MA: Addison Wesley.

Gilks, W. R., S. Richardson, and D. J. Spiegelhalter. 1996. *Markov chain Monte Carlo in practice*. New York: Chapman & Hall.

Kass, R. E., and L. Wasserman. 1996. The selection of prior distributions by formal rules. *Journal of the American Statistical Association* 91:1343–70.

Lee, P. M. 2004. *Bayesian statistics: An introduction*. 3rd ed. London: Arnold.

Park, E. S., J. Park, and T. J. Lomax. 2010. A fully Bayesian multivariate approach to before-after safety evaluation. *Accident Analysis & Prevention*, 42:1118–1127.

Park, E. S., L. R. Rilett, and C. H. Spiegelman. 2008a. A Markov chain Monte Carlo-based origin destination matrix estimator that is robust to imperfect intelligent transportation systems data. *Journal of Intelligent Transportation Systems* 12:139–55.

Park, E. S., R. Smith, T. Freeman, and C. H. Spiegelman. 2008b. A Bayesian approach for improved pavement performance prediction. *Journal of Applied Statistic* 35:1219–38.

Robert, C. P., and G. Casella. 2004. *Monte Carlo statistical methods*. 2nd ed. New York: Springer-Verlag.

Schluter, P. J., J. J. Deely, and A. J. Nicholson. 1997. Ranking and selecting motor vehicle accident sites by using a hierarchical Bayesian model. *The Statistician* 46:293–316.

CHAPTER 15

Microsimulation

15.1 INTRODUCTION

In recent years traffic microsimulation packages have become an important modeling tool for various aspects of transportation planning, design, and operations. A study in 1997 identified over fifty separate commercially available traffic microsimulation packages (Algers et al., 1997), and there is no doubt this list has grown substantially since then. Because traffic microsimulation models are becoming more widely used in a variety of transportation applications, it is important for transportation professionals to understand the concepts behind these models and analyze their output statistically.

This chapter provides a brief overview of microsimulation models, followed by the statistical methods for analyzing microsimulation output. Next, this section raises issues examined in previous chapters, such as the accuracy of the model with reference to statistical inference and the design of experiments. Methods of calibration are also briefly discussed. The final section of this chapter looks at common performance measures, and how they are calculated from simulation output.

15.2 OVERVIEW OF TRAFFIC MICROSIMULATION MODELS

Traffic microsimulation models seek to represent the interaction of the physical system (e.g., the supply: roads, intersections, traffic control, etc.) and the users (e.g., the demand: routes, driver characteristics, etc). These models are referred to as micro because they operate at an individual unit level (e.g., vehicles, people). They are referred to as simulation because they seek to model the internal processes of the system (e.g., drivers' decisions, vehicle characteristics, traffic signal operations, etc.) and not simply the output of the system. These microsimulation models are popular because the systems they represent are so complex that more traditional macroscopic models are insufficient.

Figure 15.1 provides an overview of the information flow for a generic microsimulation model. We can see that there are two basic inputs: supply (box A) and demand (box B). The supply consists of (1) the physical attributes of the network and (2) the operating strategies of the transportation agency. An example of the former would be the fact that a given intersection has a traffic signal present. An example of the latter is illustrated by that all traffic signals in the simulation model operate according to the city's timing plan. Note that the supply is typically under the direct control of the decision makers in charge of the transportation network. For example, they can add more roads, expand lanes, add transit, or change signal timing plans in the hopes of improving performance at an intersection, along a particular roadway, across a corridor, or at the system level. In contrast, the decision makers may exert considerable or almost no control over demand, depending on the situation. For example, airlines choose their prices and can manage demand by varying their fares. In contrast, highway authorities have limited ability to change prices on freeways, and consequently, they have to use other approaches, such as ramp metering or high-occupancy lanes.

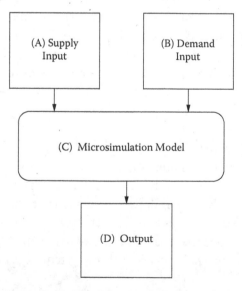

FIGURE 15.1 Schematic diagram of a microsimulation model.

The vast majority of transportation model developers find it useful to treat the physical component of the supply as a mathematical representation of nodes and links, each with its own attributes. We should note that, in general, much of transportation network supply modeling borrows heavily from mathematical graph theory, in which the nodes, or vertices, are points in space. The set of all nodes is N. The set L represents all the ordered pairs of vertices, which commonly are referred to as links, arcs, or directed edges. In essence, the links connect the nodes and hence have a direction associated with them, that is, from node a to node b. The entire network is referred to as a directed graph and is written mathematically as $G(N, L)$ (Bondy and Murty, 1976, 2008).

For the sake of convenience, the present chapter will discuss networks with respect to the roadway mode, and we will use the common transportation terminology of nodes and links. There are, however, simulation models for railways, transit lines, multimodal networks, and other transportation modes. The general principles discussed in this chapter will apply equally to these microsimulation models. The attributes of nodes typically include location coordinates—x, y, and possibly z—and type of signal control, such as uncontrolled, stop sign, or traffic signal. The nodes are often thought of as representing intersections, although this is not necessarily true from a practical point of view. Note that while the coordinates may correspond to latitude, longitude, and elevation, any spatially consistent coordinate system could be used. The links represent homogeneous sections of the roadway network, and their attributes would include number of lanes, speed limit, or grade.

Figure 15.2 shows a section of a VISSIM traffic network in Lincoln, Nebraska. The traffic network is represented by links and nodes, and this particular microsimulation network is multimodal since both roadways and railways are represented. Figure 15.3 presents a snapshot of a microsimulation from this network. This figure features a signalized intersection located near an at-grade railroad crossing. We can see that vehicles, pedestrians, trains, and traffic control can all be modeled and displayed to the user. The visualization aspects of microsimulation models make them powerful tools for communicating with decision makers and the public. However, it is important for the user to understand whether the models are reasonable, and this is the focus of this chapter.

Figure 15.4 shows a detailed view of an interchange from the Lincoln network where the nodes (circles) and links (solid lines) are shown explicitly. In this figure, the direction of movement is indicated by the arrows.

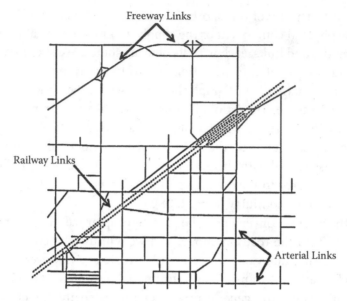

FIGURE 15.2   VISSIM subnetwork of Lincoln, Nebraska, showing streets (solid lines) and railway lines (dotted).

FIGURE 15.3   Screen shot of a microsimulation model.

It should be noted that nodes do not solely represent intersections; rather, they represent locations where the roadway characteristics change. It is also important not to assume that simply because two links cross, there is a node located at their intersection. Grade-separated road crossings, for example, would not have a node. Note that while each microsimulation model may have a different set of attributes and coding schemes, they will, for the most part, all model the physical transportation network as some form of directed graph.

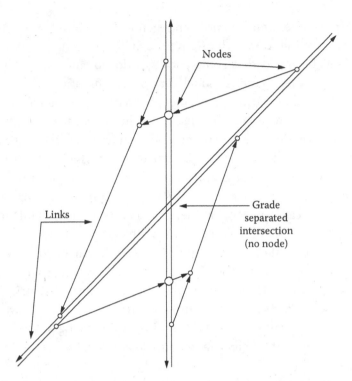

FIGURE 15.4   Directed graph of a highway interchange in Lincoln showing nodes (circles) and links (lines).

The second input (shown in box B in Figure 15.2) is the transportation demand. The demand typically takes the form of an origin-destination matrix where the number of vehicles are defined whose drivers wish to travel from a given node i to a given node j and whose drivers wish to depart their origin at some point during a specific time period. Typically, the input also includes the vehicle types (e.g., percentages of passenger cars, buses, tractor trailers, etc.), the driver types, and the respective attributes of both (acceleration capabilities, braking capabilities, perception reaction time, driving aggressiveness, etc.) associated with the demand between the two nodes. Note that the traffic demand can be static in that it does not change over the entire simulation, or dynamic in that it does change with simulation time. In addition, the more sophisticated models will define OD demand with respect to the movement of people, as opposed to vehicles, and the simulation will include a mode choice component. For example, three hundred people wish to travel from origin A to destination D, and their travel options include drive alone, carpooling, transit, and cycling. In this case the model would identify endogenously the number of vehicles of each type traveling from each origin to each destination.

It should be noted that some packages do not input an OD matrix directly. Instead, demand is modeled as consisting of (1) volumes entering the links and (2) turning movement percentages at intersections. This approach has an advantage, because the volume and turning movement information is easier to obtain in the field than from an OD matrix. While both methods can be used to represent demand, the former allows the modeler more control. For example, consider Figure 15.5, which shows a simple network with observed volumes and two potential OD matrices. The reader can easily verify that there are a large number of OD matrices that would produce the volumes on the links. Some involve significant weaving (e.g., OD Matrix B in Figure 15.5), while others do not (e.g., OD Matrix A in Figure 15.5). Clearly the amount of weaving will affect the simulation results. The authors recommend that the OD matrix be used whenever possible, as it provides more control to the user. However, the reader should be cautioned that estimating or observing large OD matrices is not a trivial problem (Nihan and Davis, 1989; Cascetta et al., 1993; van der Zijpp, 1997).

In general, the user defines a specific length of simulation time (e.g., from 8 a.m. to 10:30 a.m.) as part of the input. The microsimulation program progresses through the modeling process at small time increments, such as 0.1 seconds. In doing so, it models the interaction between the individual units, or vehicles, as they enter the network at their origin

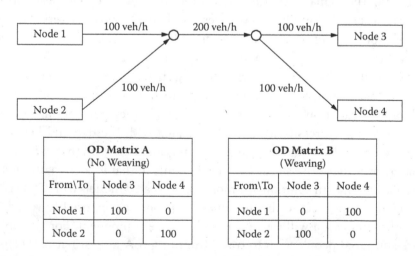

FIGURE 15.5 Simple network showing traffic volumes and potential OD matrices.

nodes, traverse it while interacting with the traffic control as well as other vehicles, and depart at their destination nodes. This process is represented by box C in Figure 15.1.

Note that the demand is typically entered as a deterministic number (veh/h) rather than as individual vehicle movements. In Figure 15.5, for example, the demand from node 1 to node 3 might be 100 veh/h. The user would input this along with attributes such as from node to node, vehicle type percentages, and driver type percentages. The use of aggregate values for input saves time and effort because the user does not have to enter specific information—from node, to node, type of vehicle, type of driver, departure time, and so forth—for 100 separate vehicles. Instead, the program translates this deterministic number into individual vehicle movements that have specific attributes that correspond to the macroscopic input. Consider the situation where one hundred vehicles per hour (veh/h) are input as traveling from node 1 to node 3 between 8 a.m. and 9 a.m. In this situation approximately one hundred vehicles will be created that have attributes associated with the input information, that is, 20% heavy vehicles and 80% passenger cars. The vehicles will then be assigned to leave at random times between 8 a.m. and 9 a.m. Based on logic that is typically internal to the simulation model, vehicles choose a route to their destination, enter the network at node 1, and traverse the network to node 3, while simultaneously interacting with both other vehicles and the transportation system. The manner of these interactions is a function of the type of vehicle and the driver type attributes. The model simulates the system at set time increments, which are typically at the decisecond (e.g. 0.1 second) interval. For example, during a particular time step, a given vehicle may move from point x to point y, a traffic signal may change from amber to red, and other changes may occur. The simulation proceeds until the end time input by the user is reached.

Modeling large cities at the decisecond level results in a considerable amount of data available for analysis. Consequently, the user typically has discretion in the information that is output (e.g., box D in Figure 15.1) as part of this process, the form it should take, and the frequency of the output. Typical output may be categorized as (1) information related to the individual vehicles statistics, (2) information related to specific links or intersections, and (3) information related to the system as a whole. In addition to aggregated data from the entire simulation, users will request output in fifteen-, thirty-, or sixty-minute increments.

It is important to emphasize that many of the phenomena in microsimulation programs, such as driver behavior, are modeled as random variables. The exact distribution of the random variable used within the model is often defined internally to the program, although the user often defines the parameters. The number of vehicles for a given OD pair, for instance, may be modeled as a normal distribution, whereas the time between arrivals (e.g., when they enter the network) may be represented as following an exponential distribution. Assume that you wanted to model the network shown in Figure 15.5 and that you input 100 veh/h traveling from node 1 to node 4 from 8 a.m. to 9 a.m. In some programs, the total number of vehicles may be random; for instance, it could be 103 veh/h or 98 veh/h. The important point, though, is that while the input values are deterministic, the model will "draw" the actual number from a predefined pdf. Similarly, the time between arrivals at which the vehicles enter the network, between 8 a.m. and 9 a.m., also will be random and would be drawn from the exponential distribution (if this pdf were chosen by the user).

Lastly, the microsimulation program developers assume the user is familiar with the assumptions underlying the logic of the model. The standard microsimulation models all have default parameters, such as a value for λ in the Poisson distribution for vehicle departure counts. As a result, the user does not necessarily need to calibrate the model before running it. Consequently, if you provide input in the proper format and run the microsimulation program, you will get output. It is then up to the reader to decide how good the simulation results are, a concept that will be explored further in the following sections.

15.3 ANALYZING MICROSIMULATION OUTPUT

Because of their flexibility and ability to model complex systems, microsimulation models have become increasingly common over the past twenty years. This section will examine the different types of statistical analyses commonly used when analyzing their output.

15.3.1 Model Calibration Validation and Verification

The terms *verification*, *validation*, and *calibration* are often used in the context of microsimulation modeling. However, these terms are not always well defined or understood. For this reason, the present section will explicitly define them, basing the definitions on common usage in the transportation engineering community.

Model verification is the process of determining if the logic that describes the underlying mechanics of the model, as specified by the model developer(s), is faithfully replicated by the model. It is important to note that model verification is not concerned with whether the logic is correct. For example, if the model developers intended that a stream of vehicles approaching an isolated intersection follow the Poisson distribution, then model verification will confirm that the modeled vehicles are indeed distributed according to a Poisson pdf. The graphical goodness-of-fit techniques described in Chapter 9 could be used for this application. The question of whether this distribution is correct is not part of model verification.

Model validation is the process of determining to what extent the model's underlying fundamental rules and relationships are able to adequately capture the targeted emergent properties of the model. As the name implies, emergent properties emerge from the model and are not defined *a priori*. In traffic microsimulation models, for instance, link capacity, density, and speed are often defined as emergent properties. Note that the emergent property might be compared to theoretical values or empirically collected data. For the capacity example, validation might involve comparing the simulated link capacity with (1) the *Highway Capacity Manual* values (HCM, 2010) or (2) observed data.

Model calibration is the process of modifying the default microsimulation parameters so that the model replicates the observed traffic conditions as accurately as possible. For example, if a Poisson distribution is deemed appropriate for modeling a given stream of vehicles arriving at a traffic signal, then the model calibration would identify the best value for the parameter λ of the Poisson distribution. A more detailed overview of model calibration will be provided later in this chapter. We should note that it is good practice to use separate data sets for the model calibration and model validation processes. In addition, the calibration step is commonly performed before the validation step.

15.3.2 Aggregate Goodness-of-Fit Measurements

A number of aggregate goodness of fit measurements can be used to quantify the degree to which the model results fit the field data. The term *aggregate* is used because all the measurements are combined into a single metric. Commonly used aggregate measures include the mean absolute error (MAE), the mean absolute proportional error (MAPE) (sometimes referred to as the Mean Absolute Error Ratio (MAER)) and the root mean squared error (RMSE), shown in Equations 15.1 to 15.3 respectively.

$$MAE = \frac{1}{N} \sum_{i=1}^{N} |O_i - E_i| \tag{15.1}$$

$$MAPE = \frac{1}{N} \sum_{i=1}^{N} \frac{|O_i - E_i|}{O_i} \tag{15.2}$$

$$RMSE = \sqrt{\frac{1}{N} \sum_{i=1}^{N} (O_i - E_i)^2} \tag{15.3}$$

where:
O_i Measure of performance (e.g., average link volume) observed from field data
E_i Measure of performance (e.g., average link volume) estimated by the simulation model
N Total number of observations (e.g., link volumes).

Each measure has its own advantages and disadvantages. As an example, the MAPE estimates error as a proportion of the observed mean and is therefore easily understood by decision makers because it has a physical meaning. For example, assume that you had measured link volumes and were comparing these to simulated link volumes. A MAPE of 0.10 would indicate that, on average, the simulation and modeled results differ by 10%. However, this measure is best suited to applications in which the denominator, the observed measure of performance, does not exhibit a wide range. If the measure of performance is link volume, to name one possibility, and the network consists of a mixed arterial-freeway network, then it is likely that the MAPE for minor roads will be larger than for major roadways. Consequently, MAPE should not be computed for all links combined, but should be computed separately for groups of links with similar observed volumes. Note that users can choose any measure of performance (link volume, occupancy, speed, etc.); however, they should indicate what metric they used when reporting the results.

The above equations are presented for a single time period. It is often valuable to apply the selected measure of performance for discrete time periods. For example, when simulating a four-hour peak period, it may be desirable to report performance measures for each one-hour or thirty-minute period so that trends may be examined over time. In this situation the aggregate performance measure is calculated for each time period. For example, the disaggregate MAE for time period t (MAE_t) is shown in Equation 15.4.

$$MAE_t = \frac{1}{N_t} \sum_{i=1}^{N_t} \left| O_{it} - E_{it} \right| \quad t = 1, \cdots T.$$

$$MAE = \frac{\sum_{t=1}^{T} MAE_t}{T} \tag{15.4}$$

where:
- t Time period t; note all time periods are of equal length
- T Number of time periods
- O_{it} Measure of performance (e.g., average link volume) observed from field data during time period t
- E_{it} Measure of performance (e.g., average link volume) estimated by the simulation model during time period t
- N_t Total number of observations (e.g., link volumes) during time period t.

Note that in transportation literature most measures of volume and demand are presented in terms of vehicles per hour (veh/h). However, by convention, this volume measurement is assumed to be based on fifteen-minute observations unless stated otherwise (HCM, 2010). Consequently, many users choose to output volume data at the 15 minute level of aggregation.

15.3.3 Statistical Analysis of Microsimulation Output

There will be many times when the user will want to know whether the simulation results are valid or whether there is a significant difference between the results in a statistical sense. This is typically determined by comparing the simulation results to one of the three sources that follow:

1. Deterministic values that are based on theory or common practice, for example, a capacity value derived from a microsimulation analysis vs. the capacity calculated from the highway capacity model.

2. Empirical data obtained from one or more field sites. The empirical performance measure output can be collected manually by the user or automatically by existing detector equipment that is part of a monitoring system (e.g., inductance loops).

3. Output from two microsimulation runs that have different supply or demand characteristics. For example, simulation 1 might use a current traffic control system, and simulation 2 an improved traffic control system. In this case, the user may be interested in determining (i) what improvements may occur from the new traffic control systems, and (ii) whether these differences are statistically significant.

Note that comparisons 1 and 2 can be components of the model validation step; that is, these types of comparisons can be used to answer the question of how well the model is performing.

In addition to identifying the point of comparison for the simulation results, the user will also have to define which performance measures will be compared. Among other measures, these may include average speed, average delay, and volumes. She will also need to define which statistical tests will be used along with their associated levels of significance. The general framework is shown in Figure 15.6.

The above decisions are not often made in isolation. For example, the question of which performance measures to use is often a function of what can be collected readily in the field (e.g., we would like roadway density but we have access only to automated occupancy measurements) and what is output from the microsimulation model (e.g., we would like approach delay by turning movement but only average vehicle delay is output from the model). Needless to say, the level of significance (or confidence) will be a function of the performance measure being estimated and the cost to obtain the necessary data. For instance, we would like 0.1% significance levels (99.9% confidence levels), but 10% significance levels (90% confidence levels) will suffice for the given application and cost of data collection.

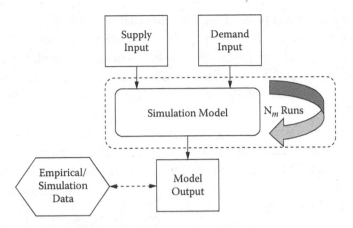

FIGURE 15.6   Statistical analysis of microsimulation results.

Importantly, by definition both the empirical and simulation results can be stochastic. With respect to the empirical data, for instance, you may measure occupancy from 7 a.m. to 8 a.m. over a period of twenty weekdays. From these measurements you can then obtain measures of central tendency (e.g., mean, median, and mode), measures of dispersion (e.g., variance), and if there is enough data, the underlying distribution.

The reason why microsimulation output may be stochastic requires a more detailed explanation. By its nature, a microsimulation model will provide the same output given the same input, all else being equal. This is an advantage from a modeling perspective because someone in Lincoln, Nebraska, can obtain the same results as someone in Chennai, India, using the exact same input data and microsimulation model. In addition, it means the microsimulation experiments or runs are repeatable—running the model with the exact same input a year later will give the same result. However, the underlying logic of traffic microsimulation models is inherently stochastic in that the models use a pseudorandom number generator to estimate the probability density functions (pdfs). These pdfs, in turn, are used to identify when vehicles depart from their origins, select vehicle types, select routes, identify behavior attributes, and so forth. The pseudorandom number generators will produce a distinct stream of more or less random numbers for different seed numbers. The seed numbers are part of the input requirements and are controlled by the user. Consequently, two different seed numbers will result in two different sets of simulation results. It is important to choose seed numbers that produce independent runs. More importantly, this means that a given scenario can be analyzed based on the average of several simulation runs using different seed values for random numbers. If a user runs the simulation N_m times with N_m different random number seeds, then measures of central tendency and dispersion (and in fact the complete cumulative distribution function) can be obtained from microsimulation models. This is shown in the dotted box in Figure 15.6.

Because of the stochastic nature of the observed and simulated data, statistical techniques are necessary for making informed decisions regarding the quality of the simulation results. For example, it is unlikely that the empirical and simulated results, such as the average speed on a specific link, will match exactly.

When a user wishes to make statements about specific physical quantities, the methods of inference discussed in Chapter 7 typically are used. A common question is to compare a stochastic variable with a deterministic value (e.g., is the average stop delay equal to or less than a theoretical value?). In this situation, simple hypothesis tests or confidence intervals (CIs), with or without bias, depending on the situation, are appropriate.

Another common question is whether the difference in a given performance measure between the observed data, denoted by Y, and the simulated data, denoted by X, is significant. In such a case, the confidence interval for the mean difference shown in Equation 15.5, which was first defined in Chapter 7, can be used.

$$\bar{X} - \bar{Y} \pm t_{df,1-\alpha/2} SE(\bar{X} - \bar{Y}) \qquad (15.5)$$

where:
X Observed measure (e.g., average travel time for the entire simulation, average speed on a specific link, etc.)
Y Corresponding simulated measure (e.g., average travel time for the entire simulation, average speed on a specific link, etc.).

As before, if the CI contains zero, then there is no significant evidence against the hypothesis that the means of the two measures are the same. We must note that X may also represent simulation results. A transportation planner, for instance, might wish to analyze the effect that the removal of a bridge may have on a system. In this case, the user could run the model with and without the bridge and then compare the results using Equation 15.5.

15.3.4 Identifying the Statistically Optimal Number of Simulation Runs

The above tests assume that the number of simulation runs, N_m, is known *a priori*. For example, you decide ahead of time that you will run the simulation twenty times and then perform the statistical analyses. However, sometimes it is useful to calculate how many simulation model runs are required for a particular analysis in order to obtain statistically significant results. The formula for answering this question is shown in Equation 15.6.

$$N_m = \left(t_{1-\alpha/2, N_m-2} \frac{\sigma_m}{\varepsilon} \right)^2 \quad m = 1, \cdots, M. \qquad (15.6)$$

where:

N_m Number of simulation runs for performance measure m

M Number of performance measures that are being considered by the user

σ_m (Estimated) standard deviation of performance measure m

$t_{1-\alpha/2, N_m-2}$ t-Statistic value for given significance level and number of simulation runs

ε Allowable error; this is often specified as a fraction of the mean value of the performance measure μ_m.

Equation 15.6 is essentially a reordering of the confidence interval bounds introduced in Chapter 7 so that the number of simulation runs is the dependent variable. It can be seen from the above equation that there are several pieces of information required to solve for N_m. The level of significance has to be defined by the analyst. Values of 10, 5, and 1% are typically chosen. In addition, the normal distribution is often used as an approximation of the t distribution when N_m is large, which means that $t_{1-\alpha/2, N_m-2}$ can be replaced with $z_{1-\alpha/2}$ on the right-hand side of the equation. More importantly, in order to solve for N_m, the standard deviation of the performance measure is required—this is problematic, as it will not be known until the traffic simulations have been run. The user is then faced with two choices: either to assume a value based on past experience or to use an iterative approach.

Step 1   Run a set number of simulations. A typical number would be on the order of ten.

Step 2   Calculate the mean and standard deviation of each performance measure, m.

Step 3   Calculate N_m for each performance measure m. The highest value is the one required for the analysis.

Step 4   Compare the number of simulation runs completed with the number identified in step 3 (i.e., compare N_m with the current number of simulation runs). Have there been enough simulation runs?

If yes, stop.

If no, run the simulation again and return to step 2.

It is a useful approximation. It produces a larger sample size than one would get if they assumed σ_m is known. It still may produce too small of a confidence interval due to the sequential choice of sample size (sequential testing was discussed in Chapter 7).

The above equation is based on the assumption that the performance measure follows the central limit theorem. For metrics that are based on average values, this is likely a valid assumption. However, in some cases, such as the variance, this might not be true and alternative methods would be required. In addition, sometimes the analyst is interested in the distribution of the performance measure, rather than just the mean. In both of these cases nonparametric tests are often used (Kim et al., 2005). For more sophisticated methods for choosing sample sizes for computer experiments, a good reference is Loeppky et al. (2009).

Lastly, it is important to note that, in general, the more disaggregate the level of analysis detail becomes, the more runs are needed. For example, if a comparison of the average travel time per link for all links and all time periods were required, it would in all likelihood involve considerably more runs than the average travel time across all OD pairs for all time periods.

15.3.5 Microsimulation Calibration

The calibration of traffic microsimulation models has received widespread attention because of the increasing use of these models for both traffic operations and transportation planning applications. The ability to accurately and efficiently model traffic flow characteristics, drivers' behavior, and traffic control operations is critical for obtaining realistic microsimulation results.

As discussed previously, calibration is defined as the process of adjusting the value of the microsimulation model parameters such that the simulation output is, as far as possible, consistent with the observed or empirical data. Because of the difficulty in collecting data in the field and the lack of readily available automatic calibration procedures, users often run traffic microsimulation models with default parameter values. Based on the authors' personal experiences, this is a bad idea—rarely do the default values result in microsimulation output that replicates empirical data. In our experience it is not uncommon to run microsimulation models using default values and obtain mean absolute error ratio (MAPE) values of well over 100%.

Figure 15.7 illustrates a conceptualization of the calibration process. The first step in calibration is the selection of one or more field sites from which field data may be obtained. The second step is to define the set of performance measures by which the model will be compared to existing field conditions. For example, the user could choose to use average speeds, travel times, average delay, or other factors. A full description of possible metrics is discussed in the last section of this chapter. Once these measures have been obtained, appropriate amounts of empirical data are collected. As discussed earlier in this chapter, the appropriate amount of data will be a function of the statistical analysis adopted and the level of confidence desired. The observed network is modeled using an initial prior microsimulation parameter set. This prior set could be based on the default values, previous calibration results, or engineering judgment. The model is run and the performance measures are then collected and estimated from the model

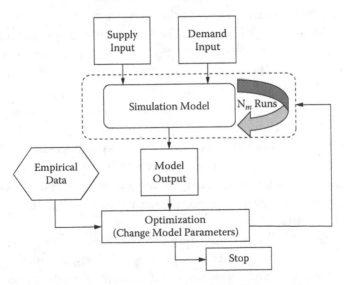

FIGURE 15.7   Overview of microsimulation calibration process.

output. Criteria are selected that define the level of similarity between the model results and the field data required for the model to be considered calibrated satisfactorily. For example, aggregate measures (e.g., MAPE), confidence intervals, and nonparametric distribution tests have all been used. Based on the results of the statistical tests, the model parameter(s) is adjusted and the process repeated. When the simulated data matches the empirical data, the model is considered calibrated and the process stops.

The calibration process outlined in Figure 15.7 is sometimes done manually with each step controlled by the user. This is particularly problematic for microsimulation model calibration because there is often a multitude of parameters (related to vehicle characteristics, driver behavior, etc.) that may require adjustment. For example, VISSIM has over thirty parameters that the user can control. Because the manual process is time-consuming, it is rarely done to any great length, and consequently only a small subset of potential parameter sets is studied. While it is better than simply using the default parameter set, the manual approach is not recommended.

15.3.6 Computer-Based Microsimulation Calibration

With the recent growth in computational resources it is now possible to develop automatic calibration procedures that mimic the process shown in Figure 15.7. One key question is how to optimize the parameter selection process. Fortunately, there are numerous optimization procedures that can be used. The gradient approach determines the search direction by evaluating the partial derivative of the objective function. A simplex algorithm is based on a geometric feature for optimizing systems containing continuous factors. A genetic algorithm (GA) is a problem-solving algorithm that emulates biological evolutionary theories to solve problems in the field of optimization. In transportation engineering, all of these procedures have advantages and disadvantages and have been used in microsimulation model calibration (Schultz and Rilett, 2005).

In the past, collecting the necessary empirical data was difficult, expensive, and time-consuming. With the recent widespread deployment of intelligent transportation systems we have access to an abundance of data on traffic systems as well as opportunities to use this data for calibration. (Rilett et al, 2001).

Once the optimization technique and data sources are selected, automating the process on a microcomputer is relatively straightforward. Of more critical importance, however, is the measure of similarity between the empirical observations and simulated results that is used in the calibration process. Many authors have used aggregated performance measures, such as average travel time or total traffic volume, in an attempt to find the best parameter set that minimizes some objective function. A common measure is the MAER, wherein the parameter set that has the lowest MAER is selected as the preferred one.

When using an aggregated performance measure, such as the average network speed, there is a tendency to assume that the parameter set that produces the lowest value of the performance measure is the best parameter set. However, this assumption is valid only when the distributions for the simulated and observed speeds are identical. In other words, the only difference between the results from different parameter sets is a measure of central tendency, such as the mean or median. Alternatively, the distribution of empirical data and simulated data can be compared using the Wilcoxon rank-sum, Kolmogorov–Smirnov, or other nonparametric tests.

Lastly, if the user is employing a statistical test in the calibration process, it is possible that he or she will have a number of parameter sets that are neither statistically different from each other nor statistically different from the empirical data (Kim et al., 2005). In this case other criteria will have to be used to choose the best parameter set from the group of statistically appropriate parameter sets. Alternatively, it is also possible that no statistically significant parameter sets are found.

It should be noted that often model calibration efforts are conducted on the sole basis of traffic volumes. Based on our experience, it is the authors' recommendation that this be avoided because it is possible to have simulated link volumes match quite closely to field traffic counts and still observe large discrepancies between modeled and actual levels of congestion. Consequently, it is recommended that if microscopic traffic simulation models are being calibrated on the basis of link volumes or turning movements, then at least one other metric—such as link speed, travel time, or queue location and length—should be used.

In practice it appears that for most calibration studies, target criteria levels are defined. One such example is ensuring that all measured link volumes are within some prespecified percentage of the observed values. Once these performance criteria are established, all reasonable attempts are made within the allotted budget and project scope constraints to reach these targets. If they are not obtained, then the level of confidence placed in the model results also must be reduced. It should be noted that there are no universally agreed upon criteria for determining when a model is calibrated. Some users simply run the calibration process a set number of times and pick the parameter set that provides the lowest MAER, and thereby the "best" result.

Hypothetically, there are three possible causes of an unsuccessful model calibration. First, because the solution space is highly nonconvex, there is no known algorithm, short of complete enumeration, that will ensure an optimal solution. As such, the results will be local optima—if the modeler is unlucky, these local optima may be fairly poor. Algorithms that attempt to search across a wide range of values, such as genetic algorithms, can help minimize this problem. Second, the calibration may be unsuccessful because an insufficient quantity of field data, or data of insufficient quality, may have been used to calibrate the parameter values. Third, the underlying logic in the microsimulation model may be inadequate to capture certain traffic behavior correctly; thus, it would be impossible to correctly calibrate the model. Unfortunately, when a user has difficulty calibrating a model, it is generally not clear which of these possible causes, or combination of possible causes, is the source.

15.3.7 Absolute vs. Relative Model Accuracy

When considering which of the many traffic microsimulation packages to use, a key question—and, unfortunately, one that is often overlooked—is that of identifying the intended application. In transportation engineering microsimulation models have been used for many applications, as demonstrated in the following list.

Traffic operations	Selecting optimal signal timing plans, analyzing the capacity of roadways, and calculating travel times
Roadway design	Identifying the feasibility of different designs and analyzing merge conditions
Evaluation	Estimating performance measures such as average delay and fuel emissions
Transportation planning	Analyzing different investment scenarios: roadway vs. light rail transit

Identifying the application is important for a number of reasons. Intuitively, different traffic microsimulation models will be preferred depending on the application. The application will also define which statistical approaches (discussed in this textbook) will be required, along with their level of sophistication. For example, if the application requires that a specific performance measure be provided, such as intersection control delay, then the model must offer accurate estimates of this parameter. In

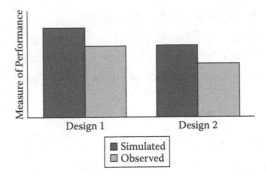

FIGURE 15.8 Absolute vs. relative model reliability.

addition, the modeler must be able to identify the accuracy of the estimate, as was discussed earlier in this chapter.

However, if the goal of the model is to identify the best design alternative, then the absolute accuracy of the model is less of an issue provided that the performance measures generated from the model correctly reflect the differences between the designs. This is illustrated in Figure 15.8 for designs 1 and 2 where the lower the value of the performance measure (e.g., delay) the better the design. For both designs, the model overestimates the measure of performance, but it does correctly reflect the relative benefits associated with design 2.

Even for the most sophisticated microsimulation models, there is considerable difficulty in identifying "ground truth" from empirical data. It is rarely possible to know within well-defined statistical limits the origin-destination traffic matrix (e.g., the demand) for the current time period. However, it is possible to estimate these demands such that when applied to the road network, the resulting traffic volumes reasonably reflect observed traffic volumes. As we discussed in this chapter, this may or may not be sufficient for a given application. For instance, matching link volumes correctly does not ensure accurate representation of link travel times or delay.

15.4 PERFORMANCE MEASURES

Transportation engineers use performance measures to analyze the transportation system. As we discussed in Section 15.3, performance measures are often used in the calibration process and hypotheses testing, and may be calculated from observed or simulated data. This section provides a brief overview of performance measures followed by a description of those most commonly used.

15.4.1 Overview of Performance Measures

Performance measures are broadly used for simplification, quantification, and communication (Zietsman and Rilett, 2002b). The following are some of the most important applications of performance measures:

1. Provide a broad perspective

2. Assess facility or system performance

3. Calibrate models

4. Identify problems

5. Develop and assess improvements

6. Formulate programs and priorities

7. Educate a wide range of interest groups

8. Set policies.

Performance measures are typically used to provide the decision maker with the quantitative information necessary to make and monitor decisions. Typical levels of aggregation of performance measures as well as examples of each type of metric are shown in Table 15.1 (U.S. D.O.T. 1996, Poister, 1997, Zietsman and Rilett, 2002b). It should be noted in this table that the two highest levels of aggregation, namely, goals and objectives, are direct products of strategic planning exercises. The two lowest levels,

TABLE 15.1   Levels of Aggregation

Level of Performance Measurement	Types of Performance Measures Required	Examples of Relevant Measures
Goals	Overall goal for sustainable transportation	To have a sustainable transportation system
Objectives	Social, environmental, and economic objectives	To have a safe road-based transportation system
Indices	Aggregated or integrated performance measures	Safety index
Performance measures	Input, output, or outcome measures	Fatalities per 100 million miles of travel
Information	Manipulated data	Vehicle miles of travel and number of fatalities
Data	Raw data	Volume counts and accident records

information and data, are derived from operational management and data collection. This chapter focuses on the performance measures that are typically developed from simulation models, such as Paramics or VISSUM.

Table 15.2 includes a summary of the attributes of a good performance measure (Meyer and Miller 2000, OECD 2001, Zietsman and Rilett 2002b). Notice that the fifteen attributes of a good performance measure suggested in Table 15.2 are effectually a wish list that the transportation engineer strives toward. It will be relatively rare for a performance measure to possess most of these attributes. In some instances certain attributes of a good performance measure are not compatible, and a particular

TABLE 15.2   Attributes of a Good Performance Measure

Quality	Explanation
1. Able to discriminate	Must be able to differentiate between the individual components that are affecting the performance of the system
2. Able to integrate	Must be able to integrate environmental, social, and economic aspects
3. Acceptable	The general community must assist in identifying and developing the performance measures
4. Accurate	Must be based on accurate information, of known quality and origin
5. Affordable	Must be based on readily available data or data that can be obtained at a reasonable cost
6. Appropriate level of detail	Must be specified and used at the levels of detail and aggregation appropriate for the questions it is supposed to answer
7. Have a target	Must have a target level or benchmark against which to compare it
8. Measurable	The data must be available, and the tools need to exist to perform the required calculations
9. Multidimensional	Must be able to be used over time frames, in different geographic areas, with different scales of aggregation, and in the context of multimodal issues
10. Not influenced	Must not be influenced by exogenous factors that are difficult to control or of which the engineer is not even aware
11. Relevant	Must be compatible with overall goals and objectives
12. Sensitive	Must detect a certain level of change that occurs in the transportation system
13. Show trends	Must be able to show trends over time and provide early warnings about problems and irreversible trends
14. Timely	Must be based on timely information that is capable of being updated at regular intervals
15. Understandable	Must be understandable and easy to interpret, even by the community at large

performance measure will therefore not comply with all characteristics. As an example, it is very difficult for a performance measure to be simple (i.e., understandable at the community level) and at the same time able to address certain complex multidimensional aspects. Therefore, it is often necessary to have a variety of indicators for the different applications. Because of the wide range of potential performance measures for all possible applications, the nature of this chapter will be more technical. That is, it will focus on measures relevant to traffic and transportation engineers who are interested in quantifying the performance of the transportation system.

To date, performance measures have been related more toward the operational aspects of transportation systems. This is because transportation decision makers have tended to concentrate on enhancing the supply side of transportation. Recent legislation in the United States has demanded a paradigm shift in terms of performance measures. For example, mobility, accessibility, efficiency, and effectiveness have all recently become more important for selecting performance measures (Turner et al., 1996).

15.4.2 Common Performance Measures Used in Simulation Analysis

The output from the simulation model is used to quantify performance measures, which can be categorized into three types:

1. Point measures (i.e., spot mean speed, throughput)

2. Link-based measures (i.e., link travel time, space mean speed, delay)

3. Corridor/system measures (i.e., average vehicle speed, average person delay).

Typically, the majority of data that are collected automatically in the field consist of point measures because of the nature of the technology available. For example, loop detectors, video detection, and radar, which are among the most commonly used traffic collection devices, all provide traffic information obtained at a single point, by placing them throughout the network. Electronic toll tags and cell phone technology allow more vehicles to be tracked both spatially and temporally as they traverse the network. Obviously, as these systems become more widespread, link-based data will become more readily available.

It should be noted that most simulation models can provide information for all three categories. In general, the second category is most useful because it allows the analyst to study spatial and temporal trends. Regardless of the type of data chosen, the amount of output can be overwhelming for large simulations. Consequently, many analysts focus on aggregate measures, at least for the preliminary analysis.

Often traffic engineers are interested in performance measures related to congestion, which may be classified by four dimensions (Shrank and Lomax, 2009):

1. Duration (i.e., length of time the facility is affected by congestion)

2. Extent (i.e., number of people affected and the geographic location)

3. Intensity (i.e., severity of the congestion; typically, it is actual trip experience in relation to expected trip experience)

4. Reliability (i.e., variation of the first three dimensions).

In general, the preferred performance measures are used to measure congestion variation based on the levels of analysis and usage of the measure. Approach delay, person delay, and similar performance measures are useful for specific intersections and locations. Performance measures such as travel time, speed, and delay per vehicle are useful up to the corridor level. At high levels of analysis, cumulative measures are often used. In addition, congestion indices are used occasionally to help quantify congestion because of the difficulty in interpreting relative conditions. The following sections will outline the most common performance measures used in microsimulations.

15.4.3 Travel Time

Travel time is a very important performance measure for analyzing traffic operations. It can be measured on a link, a corridor, an origin-destination (OD), or system basis. It may also be analyzed by time period or for the entire day. The travel time for a given vehicle on a link is shown in Equation 15.7:

$$t_{ia} = t_{ia}^x - t_{ia}^e \quad a = 1, \cdots, A; i = 1, \cdots, N. \tag{15.7}$$

where:

t_{ia} Travel time on link a for vehicle i

t_{ia}^x Time vehicle i exited link a

t_{ia}^e Time vehicle i entered link a

N Number of vehicles being simulated
A Number of links in network
e Enter
X Exit

The average travel time for link a for a given period, p, is shown in Equation 15.8. Note that in order to utilize Equation 15.9, the time period has to be defined. It is important to note that for nonlinear aggregate relationships, such as aggregate-based air pollution estimate techniques, the selection of the time period can affect the results, all else being equal. For example, if the travel time is input to a nonlinear model, then the results will be a function of the time period chosen. Assume that you ran a simulation model and obtained average speed data, volume counts, vehicle types, and so forth on a minute-by-minute basis. It is possible, and indeed probable, that if you calculated the amount of pollution produced over a thirty-minute period and then calculated the total amount of pollution over two fifteen-minute periods, using the same data, you would obtain different estimation amounts. This outcome is true even though the simulated data were the same in both situations (Zietsman and Rilett, 2002b).

Due to the nature of simulation models, it is important to identify the procedure for recognizing whether a vehicle belongs to a given period, p. A typical approach would be to identify all vehicles that leave, or enter, the link during period p as belonging to that period. More sophisticated approaches attempt to weigh the effect of a vehicle on a given link for two or more time periods. Intuitively, as the period length increases, the above effect diminishes. In addition, as link length or congestion increases, this effect increases. We should also note that for most users this decision will be made by the model developers. For example, link travel time statistics for a given period may only apply to vehicles that departed the link during the specified time period. It is the user's responsibility to understand how the output is calculated.

$$N_{ap} = \sum_{i=1}^{N} E_{iap} \qquad a=1,\cdots,A; \quad p=1,\cdots,P \qquad (15.8)$$

$$\bar{t}_{ap} = \frac{\sum_{i=1}^{N_p} t_{ia} E_{iap}}{N_{ap}} \qquad \forall i \in p, \quad p=1,\cdots,P, a \in A \qquad (15.9)$$

where:

$\bar{t}_{ap}$ Mean travel time on link a during time period p

N_{ap} Number of vehicles associated with (i.e., entering, exiting, or on) link a during time period p

E_{iap} Indicator variable = 1 if vehicle i is associated with link a during time period p

P Number of time periods.

The corridor (OD movement) travel time can be calculated in a manner similar to that of Equation 15.9, where each vehicle is defined with respect to its entry and exit of the corridor (OD movement). As the corridor or OD travel time increases, the process of identifying which vehicle belongs to which time period, entry or exit, becomes more critical.

There are two important points to note about the above formulas. First, while there have been many empirical travel time studies over the years, of which the Texas Transportation Institute's Mobility Study is arguably the most famous, many do not actually measure travel times (Schrank and Lomax, 2009). Rather, they measure spot speeds at various inductance loop detector locations and extrapolate travel times from these measurements. It is relatively easy to show that using instantaneous speeds (or spot speeds) leads to biased estimates of travel time (or space mean speeds). While corrections are available, they are rarely applied in practice.

The second key point to remember is that the average corridor (OD) travel time in any given period is *not* simply the sum of the average travel times across all links that compromise the corridor (OD) movement. This approach would be an approximation, and the bias would be a function of corridor length, link travel times, travel time variability, as well as other factors. Furthermore, the error is most severe for the variability estimate in comparison to the mean travel time estimate.

15.4.4 Total Delay

Total delay is defined as the sum of time lost due to congestion, traffic control devices, and other factors. It may be determined quantitatively by the difference between the actual travel time and a base-level travel time of all the vehicles in a given time period. There are numerous base levels that could be chosen. However, because delay is often used at traffic signals on an approach lane basis, the free-flow travel time is typically used. Note that this free-flow time is calculated based on the assumptions that there are no traffic signal interruptions and no competing volume. This has numerous advantages, such as the fact that other baseline values may change during the simulation. For example, suppose you choose a baseline measuring the travel time associated with level of service C. If the traffic signal timing changes during the simulation, the definition of level of service C may also change. Total delay for a given vehicle is shown in Equation 15.10.

$$d_{ia} = t_{ia} - t_a^f \qquad i = 1, \cdots, N; \quad a = 1, \cdots, A \tag{15.10}$$

where:

d_{ia} Travel delay on link a for vehicle i

t_a^f Baseline travel time (for the interchange analysis a free-flow travel time is recommended).

The average delay on a given time period is calculated using Equation 15.11. Similar to the travel time calculation, this calculation requires that the researcher develop a method for determining which vehicle belongs to which period. As before, a typical approach would be to choose a common rule, such as vehicle i exits the link during period p. Regardless of the rule the user chooses, it is vital that it is the same across all performance measures.

$$\bar{d}_{ap} = \frac{\displaystyle\sum_{i=1}^{N} d_{ia} E_{iap}}{N_{ap}} \qquad a = 1, \cdots, A; \quad p = 1, \cdots, P \tag{15.11}$$

where:

$\bar{d}_{ap}$ Average delay on link a during period p.

Typically, delay is estimated over the entire network. Equations 15.12 and 15.13 give the formulas for calculating the time period and total network estimates, respectively.

$$\bar{d}_p = \frac{\displaystyle\sum_{a=1}^{A} \sum_{i=1}^{N} d_{ia} E_{iap}}{\displaystyle\sum_{a=1}^{A} N_{ap}} \qquad p = 1, \cdots, P \tag{15.12}$$

where:

$\bar{d}_p$ Average delay on network during time period p

$$\bar{d} = \frac{\sum\limits_{p=1}^{P} \sum\limits_{a=1}^{A} \sum\limits_{i=1}^{N} d_{ia}}{\sum\limits_{p=1}^{N} \sum\limits_{a=1}^{A} N_{ap}} \qquad (15.13)$$

where:

$\bar{d}$ Average delay on network.

Note that delay is often used at intersections to compare the effect of different control strategies. In this case, the delay is calculated based on the links that enter a given intersection as shown in Equation 15.11.

$$\bar{d}_{bp} = \frac{\sum\limits_{a=1}^{A} \sum\limits_{i=1}^{N} B_a d_{ia} E_{iap}}{\sum\limits_{a=1}^{A} B_a N_{ap}} \qquad p=1,\cdots,N, \quad b=1,\cdots,B \qquad (15.14)$$

where:

$\bar{d}_{bp}$ Average delay at intersection b during time period p
B_a Indicator variable; equal to 1 if link a exits to intersection b and 0 otherwise
B Number of intersections on the network.

The average intersection delay for the entire simulation period for each intersection b can be calculated by summing the results of Equation 15.7 over all periods and dividing by N. Traffic engineers will sometimes calculate delay on an approach, as opposed to a lane, basis.

At this point it is pertinent to explain that delay and travel time are naturally correlated. This is typically because neither will provide as much information individually as they do together. For example, an average travel time of thirty minutes or an average delay of twenty minutes has different meanings depending on location (Houston, Texas, or Omaha, Nebraska) and time of day (peak or off-peak period). Together, however, they provide information on the relative effect of congestion and traffic control on mobility.

15.4.5 Queue Length (Oversaturation)

When judging the operational quality of a traffic network, particularly near intersections and on and off ramps, a key performance measure is queue length. This measure is readily available from most traffic micro-simulation models. The average queue length is provided, but traffic engineers are often more interested in the maximum queue length and/or—the amount of time by which a given queue is greater than some set level. The key question is whether queues spill back and affect traffic operations at other locations. For instance, a queue at a traffic signal may affect an upstream traffic signal, or a ramp queue may affect operations on freeway main lane operations. It may be assumed (incorrectly), for example, that intersection operations do not affect one another. One way to test this assumption would be to check whether the maximum queue ever spills back toward the upstream links in such a manner as to potentially affect operations at the upstream intersection.

While queue length statistics are often output from simulation models, it is not always clear how these statistics are derived. For this reason, comparing simulated and observed queuing statistics is often problematic because there are so many different methods of defining when a vehicle is actually in a queue (e.g., does it have to be stopped or below some threshold speed?). While a complete exposition of the topic is beyond the scope of this textbook, this point does reiterate the importance of comparing like with like. If the empirical performance measures, or statistics, are calculated in a different manner than the simulated performance measures, then direct comparisons are difficult and can lead to erroneous conclusions.

Traffic engineers often are interested in the amount of double cycling—the number of vehicles that wait for more than one cycle at a traffic signal before being allowed to proceed. This figure is of consequence for traffic engineers because it is a direct measure of oversaturated conditions, that is, when demand is greater than supply. However, this value is rarely output by traffic microsimulation models, and therefore queue length is often used as a surrogate measure. Intuitively, queue length is related to travel time and delay. More importantly, the performance measure does provide information on why delays may be occurring and, more importantly, where the critical delays are located. For example, the information could be used to identify potential locations for improved traffic signal operation or geometric improvements.

15.4.6 Number of Vehicles Completing Simulation

As described in the preceding two sections, it is important to define, *a priori*, how a given travel time or delay measure will be counted (i.e., when a vehicle exits or enters a link, or based on some weighted average). It is equally important, however, to make sure that any comparisons made between competing scenarios are valid. As an example, simulation 1 might have better average delay values than simulation 2. However, if simulation 2 has a greater throughput (i.e., more vehicles arrive at their destination), then it may not be clear which is the more desirable alternative. Therefore, an important performance measure for simulation programs is the number of vehicles completing their trips. Note that this is one reason why average travel time and delay values are often recommended rather than total travel time or delay. If all simulations had the same number of vehicles completing their trips, then the total travel time (delay) and average travel time (delay) would provide the same information.

There are two important points about this performance measure to bear in mind. First, if the demand is for a given time period, from 8 a.m. to 9 a.m., and the simulation runs until the end of that time period, finishing at 9 a.m., there will always be vehicles remaining on the network when the simulation ends. Second, because of the stochastic nature of traffic microsimulation models, it is unlikely that each simulation run will have the same number of vehicles completing their trips. The analyst then has two choices. One is to simply calculate this metric and use it in conjunction with the other metrics during the decision-making process. Alternatively, a base condition, such as 95% of OD movements completed, can be used whereby any simulation that does not meet this criterion is either (1) rejected from further consideration or (2) adjusted (i.e., traffic signal control, or traffic operation control) so that the OD movements are satisfied.

15.4.7 Percentage of Corridor (OD Movements) Congested (PCC) Measure

Often there are particular OD movements or corridors that are of importance to the analyst and require additional insight. For example, a transportation agency might be concerned with highway mobility, and therefore detailed metrics related to highways are expected to be important. As discussed previously, traffic delay is often used as a metric for analyzing traffic control at intersections. For freeway corridors, a common metric is the percentage of the corridor that is congested.

Congestion can be defined subjectively in terms of unacceptable speeds, travel times, or delay. It is recommended that congested travel be defined in terms of space mean speed (i.e., travel time), as the traveling public and decision makers have an intuitive sense of this metric. In addition, a base condition is required to which all measures are applied. Once the appropriate congestion metric and base conditions are identified, the PCC performance measure can be calculated as shown in Equation 15.15.

$$D_{ap} = \begin{pmatrix} 1 \text{ if } \dfrac{L_a}{t_{ap}} < S_a \\ \\ 0 \text{ otherwise} \end{pmatrix} \quad p = 1, \cdots, P; a = 1, \cdots, A$$

(15.15)

$$P_{cp} = \dfrac{\displaystyle\sum_{a=1}^{A} C_a D_{ap}}{\displaystyle\sum_{a=1}^{A} C_a} \quad p = 1, \cdots, P; c = 1, \cdots, C$$

where:

P_{cp} Percentage of time corridor C is congested during period p; note that for this equation space mean speed is chosen, but other metrics can be readily applied

L_a Length of link a

S_a Baseline metric (i.e., space mean speed) on link a

D_{ap} Indicator variable; equal to 1 if link a is congested (i.e., below baseline metric for link a) and 0 otherwise

C_a Indicator variable; equal to 1 if link a is part of corridor c and 0 otherwise

C Number of corridors on the network.

Note that the corridor performance measure, p_{cp}, is only defined with respect to links. This is problematic from two perspectives. First, the links on the network often have different lengths. Therefore, if one were to take a congested link of 2 kilometers length and split it into two, the congestion corridor metric would increase even though the input data would be essentially the same. Second, vehicles might experience below-threshold speeds on sections of a given link but still have an average space mean speed higher than the baseline value. These values would not be calculated in the metric. Therefore, some analysts prefer to calculate speed in small blocks, say, 100 meters, and then aggregate to get the percent congested performance measure. This is relatively easy to do in the new simulation models, such as TRANSIMS (Nagel et al, 1998). However, in most traffic microsimulation models obtaining this data is considerably more difficult. Consequently, if these types of metrics are chosen, it is recommended that the link lengths be kept approximately the same size. Alternatively, Equation 15.12 can be weighted by link length or volume to indicate the importance of the congestion on a given link.

15.4.8 Travel Time Variability

The reliability of travel time is a very important sustainable transportation performance measure. The standard deviation of travel time, which is calculated using Equation 15.16, is a measure of dispersion of the individual travel times relative to the mean travel time. In general, the lower this value is, the less variability there is in travel times and speed, and the more reliable the system becomes.

$$\bar{\sigma}_{ap} = \sqrt{\frac{\sum_{i=1}^{N} E_{iap}\left(t_{ia} - \bar{t}_{ap}\right)^2}{N_{ap}}} \quad \forall \ p=1,\cdots,P; \ a \in A \qquad (15.16)$$

where:

$\bar{\sigma}_{ap}$ Standard deviation of travel time on link a during time period p.

As before, the value of the standard deviation in and of itself is not that useful from a purely analytical perspective. A standard deviation of ten minutes on a link, for example, with a mean travel time of ten minutes would be considered less reliable than a link with a standard deviation of ten minutes on a link with a mean travel time of one hundred minutes. Often the coefficient of variation, which can be computed with Equation 15.17, is used to give a measure of its importance.

$$CV_{ap} = \frac{\bar{\sigma}_{ap}}{\bar{t}_{ap}} \quad \forall \ p=1,\cdots,P; \ a \in A \qquad (15.17)$$

where:

CV_{ap} The coefficient of variation of travel time on link a during time period p.

Similar to the travel time metric, the standard deviation output of many links is difficult to examine, particularly for large networks. Consequently, this information is often aggregated into corridor, OD, or total values. The standard deviation for all trips on a network is shown in Equation 15.18.

$$\bar{\sigma}_{ap} = \sqrt{\frac{\sum_{i=1}^{N}\left(t_i - \bar{t}\right)^2}{N}} \tag{15.18}$$

where:

$\bar{\sigma}_{ap}$ Standard deviation of travel time for all completed trip movements

$\bar{t}$ Average travel time for all completed trip movements

t_i Completed travel time for vehicle i.

15.4.9 Level of Service

In the *Highway Capacity Manual*, density is used to define levels of service (LOSs) for freeway sections. Density is defined as the number of vehicles occupying a given length of a lane or roadway at a particular instant. It can be computed with Equation 15.19 (Prassas et al., 2004). The maximum densities associated with levels of service A, B, C, D, E, and F are 6, 10, 15, 20, 29, and greater than 29 passenger cars per kilometer per lane, respectively.

$$D_{at} = \frac{Q_{at}}{L_a} \qquad \forall\ t=1,\cdots,T;\quad a=1,\cdots,A \tag{15.19}$$

where:

D_{at} Density on link a at time t

Q_{at} Vehicles on link a at time t.

If the flow rate and space mean speed are available, they can be used to calculate density as well. Because of the one-to-one relationship between density and the former variables, it is often used as a baseline performance measure. The authors caution against these types of LOS metrics because (1) the input variables (e.g., speed, flow, etc.) provide more information than the LOS measure, and (2) LOSs are highly correlated to other metrics—such as the percent of the corridor congested—and consequently are of limited use.

15.4.10 Travel Rate

Over the years there have been a number of other performance measures identified for freeway corridors. One is the travel rate, which is the rate of motion in minutes per kilometer for a specified roadway segment or vehicle trip. It is calculated by dividing the segment travel time by the segment length, as shown in Equation 15.20.

$$r_{ap} = \frac{\bar{t}_{ap}}{L_a} \quad \forall a = 1, \cdots, A; \quad p = 1, \cdots, P \quad (15.20)$$

where:

r_{ap} Travel rate on link a during time period p.

As the above formula shows, the travel rate is simply the inverse of space mean speed modified by the link length. Given that space mean speed is better understood by most decision makers, the authors recommend its use.

15.4.11 Performance Indices

Many indices have been developed over the years that attempt, in a single metric, to define how a given system is performing. The most visible use has been in congestion studies in the United States (Schrank and Lomax, 2009). While useful for providing a broad and aggregated overview of network performance, they are not readily applicable for model validation or calibration.

15.5 CONCLUDING REMARKS

This chapter examined transportation microsimulation models that are fast becoming the modeling system of choice for large-scale traffic and planning studies. The first part of the chapter provided a brief overview of the general structure of microsimulation models. The focus was on the supply input, the demand input, and the microsimulation output. It was demonstrated that these models are, by definition, stochastic and probabilistic in nature. Consequently, the statistical approaches developed in this textbook can be utilized.

The second part of the chapter focused on the methods for comparing microsimulation output with (1) theoretical values, (2) empirical data, and (3) other microsimulation data. These comparisons could be used for model validation for quantifying the effects of proposed network improvements. A methodology for calculating the number of simulation runs was also provided so that statistically meaningful comparisons could be made. This section discussed model calibration in detail. Lastly, the most common metrics used in microsimulation models were presented and discussed with respect to their mathematical meaning and their respective formulas.

HOMEWORK PROBLEMS

1. A microsimulation model was run in which the vehicles entering the network over the span of an hour were assigned an entry time based on a pdf. Three sets of data are provided that have a mean of 300 veh/h. How many vehicles are created in each data set? Why is the number of vehicles not equal to 300? Verify that the data is Poisson, normal, or uniformly distributed, and estimate their respective parameters.

2. A freeway in Omaha was modeled using a microsimulation package. The model was calibrated and then run ten times. Measured and simulated values are provided. Is there a statistical difference between the microsimulation results and the empirical data?

3. Repeat question 2 for the systemwide performance measures that are provided in the data side. Is there a difference in results?

4. A local traffic engineer wants to understand the effect of a bridge outage on the system. He runs the model ten times with the bridge in the network and without. Using the data provided, identify if there was a statistically significant difference between the simulation runs.

REFERENCES

Algers, S., E. Bernauer, M. Boero, L. Breheret, C. Di Taranto, M. Dougherty, K. Fox, and J.-F. Gabard. 1997. *Review of micro-simulation models.* SMARTEST Project Deliverable 3. Brussels: European Commission DGVII.

Bondy, J. A., and U. S. R. Murty. 1976. *Graph theory with applications.* New York: American Elsevier Publishing Company.

Bondy, J. A., and U. S. R. Murty. 2008. *Graph theory.* Graduate Texts in Mathematics 244. New York: Springer.

Cascetta, E., D. Inaudi, and G. Marquis. 1993. Dynamic estimators of origin-destination matrices using traffic counts. *Transportation Science* 27:363–73.

Eisele, W. L., L. R. Rilett, K. Mhoon, and C. Spiegelman. 2001. Using intelligent transportation systems (ITS) travel time data for multi-modal analyses and system monitoring. *Transportation Research Record* 1768:148–56.

Highway capacity manual. 2010. Washington, DC: Transportation Research Board, 2010.

Kim, W., S. K. Kim, and L. R. Rilett. 2005. Calibration of micro-simulation models using non-parametric statistical techniques. *Transportation Research Record* 1935:111–19.

Loeppky, J. L., J. Sacks, and W. J. Welch. 2010. Choosing the sample size of a computer experiment: A practical guide. *Technometrics* 366–76.

Meyer, M., and E. J. Miller, *Urban transportation planning*, 2nd Edition, McGraw-Hill, New York: New York, 2000.

Nagel, K. P. Stretz, M. Pieck, S. Lecky, R. Donnelly, and C. L. Barrett, "TRANSIMS Traffic Flow Characteristics," Los Alamos Nation Laboratory, Report LA-UR-97-3530, July 1997.

National Transportation Systems Performance Measures. Report DOT-T-97-04, U.S. Department of Transportation, Washington, D.C., April 1996.

Nihan, N. L., and G. Davis. 1989. Application of prediction-error minimization and maximum likelihood to estimate intersection O-D matrices from traffic counts. *Transportation Science* 23:77–90.

OECD. 2001. *Performance indicators for the road sector*. Organization for Economic Cooperation and Development.

Poister, T. H. Performance measures in state departments of transportation, national cooperative highway research program, Synthesis of Practice, No. 238, Transportation research board, National Research Council, Washington, D.C., 1997.

Prassas, E. S., R. P. Roess, and W. R. McShane. 2004. *Traffic engineering*. Englewood Cliffs, NJ: Prentice-Hall.

Schrank, D., and T. Lomax. 2009. *2009 urban mobility report, Texas Transportation Institute*. College Station: Texas A&M University System.

Schultz, G., and L. R. Rilett. 2006. Calibrating CORSIM commercial motor vehicle distributions. *Transportation Research Record* 1934:246–55.

Turner, S., M. E. Best, and D. L. Schrank. 1996. *Measures of effectiveness for major investment studies*. Southwest Region University Transportation Center, Texas Transportation Institute. College Station, TX.

Van der Zijpp, N. J. 1997. Dynamic OD matrix estimation from traffic counts and automated vehicle identification data. *Transportation Research Record* 1607:54–64.

Zietsman, J., and L. R. Rilett. 2002a. Analysis of aggregation effects in vehicular emission estimation. *Transportation Research Record* 1750:56–63.

Zietsman, J., and L. R. Rilett. 2002b. *Sustainable transportation: Conceptualization and performance measures*. Report SWUTC/02/167403-1. Southwest Region University Transportation Center, Texas Transportation Institute.

Appendix: Soft Modeling and Nonparametric Model Building

The models that have been discussed in this textbook, such as the normal distribution and the straight-line regression, are very powerful tools in transportation engineering. For example, using regression we can predict the effect of freeway traffic given the traffic on arterial roads. Using normal distribution theory, we can predict the percentage of observations that will fall with one, two, or three standard deviations of the mean.

Average speeds are often reasonably modeled as normally distributed, and traffic congestion can often be modeled as growing at a linear rate (at least until capacity is reached). However, there are many situations, such as modeling average speeds over time and modeling national congestion levels, where a normal probability distribution model and a straight-line model may not lead to acceptable results. For example, from a strictly theoretical perspective, the change in average speeds over time may be normally distributed, but where the mean and standard deviation have a temporal component or where there are occasional loop detector failures that produce outliers. Sometimes polynomial regression models fail to provide adequate models. In these situations, methods that are more robust are required. That is the subject of this appendix. These techniques should be used when the more traditional models do not provide adequate results or when the assumptions underlying the traditional models cannot be met.

To further motivate this discussion, consider Figure A.1, which shows a histogram of observed speeds where the mean is 52 mph and the standard deviation is 13 mph.

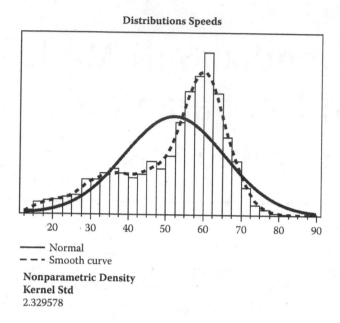

Distributions Speeds

——— Normal
– – – Smooth curve

Nonparametric Density
Kernel Std
2.329578

FIGURE A.1   Houston speeds combined: A smoothed histogram.

If the analyst makes the assumption that the data may be modeled as a normal distribution (with a mean of approximately 52 mph and a standard deviation of approximately 13 mph), the resulting model may be plotted as a solid-line curve, as can be observed from Figure A.1. It may be seen that the normal model clearly underestimates the modal speeds. Remember that a chi-square test or goodness of fit test may also be used to test whether the distributions are similar.

An alternative approach is to use a smoothed density function to represent the random variable. This is shown as a dotted line, and it may be seen that it more accurately summarizes the graphed data. While the normal curve has only one peak or mode, at the sample mean, the smoothed density has its main peak at 65 mph, and smaller peaks at 30 and 40 mph.

The travel time index for Nashville-Davidson, Tennessee, is plotted from 1982 to 2003 in Figure A.2. It may be seen that the scatter plot shows a nonlinear trend. Superimposed on the graph are the least-squares fitted straight-line, fourth-degree polynomial fit, as well as a nonparametric smoothing spline. While the fourth-degree polynomial and spline fit the plotted data well, it is clear that extrapolations using the polynomial to future years would likely be intolerably inaccurate. This poses a fairly standard statistical question in transportation engineering. Is it

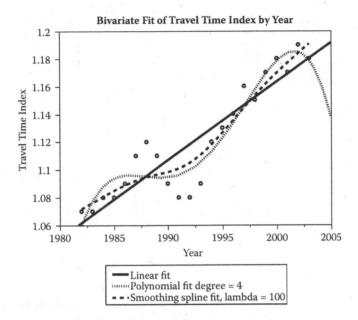

FIGURE A.2   An example of a smoothed scatter plot of travel time vs. year.

more important to choose a model that has less bias or a model that has smaller variance? The reality is there is no one answer, because each situation is different. However, the goal of this appendix is to teach analysts how to identify the trade-offs so they can choose the best answer for their application.

In this appendix we present methods for soft modeling of densities and regression functions. The methods are known to work well when the choice of a parametric model such as a normal density or straight line is not acceptable. In these situations the soft modeling or smoothing methods typically provide less biased descriptions and predictions than the parametric models do. The trade-off is that they often have a somewhat larger variance than fitted parametric models. So the transportation professional often has a choice of choosing a nonparametric estimator with less bias and a somewhat bigger variance or a parametric estimator with a bigger bias but a smaller variance. The best choice of estimator varies from data set to data set and problem to problem.

We start this discussion by focusing on kernel estimates. As an aside, if the sample is large, this approach is nearly equivalent to using smoothing splines (used by JMP) and nearest neighbor estimates (see Simonoff (1996) or Silverman (1986) for more detail, definitions, and extensive

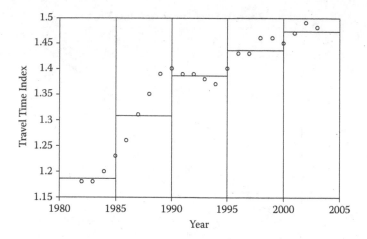

FIGURE A.3   An example of a crude soft modeling method.

reference lists). In fact, as a practical matter, most smoothing methods in practice for many problems give similar solutions and estimates. In order to motivate the approach, consider the scatter plot in Figure A.3, which shows the travel time index vs. year.

In Figure A.3, the mean of the data belonging to each five-year partition is drawn as a horizontal line. In some sense estimating the mean curve by using piece-wise averages is a model-free or nonparametric approach. Note that a number of questions arise almost immediately with this approach. We might wonder why the five-year intervals start at 1980 (instead of 1978 or 1983) and why use five-year bin sizes (instead of four or six years)? It is important to note that the discontinuity of the piece-wise averages at the boundaries is not desirable in most transportation applications. These questions will be addressed in the coming sections. However, the point of Figure A.3 is that the idea of local averaging is the key to nonparametric estimation.

Intuitively there are a number of ways to improving the above approach. The first way is to average those observations within 2.5 years of the date for which we want our travel time index. In that way we would be estimating in the middle of our local region. Second, in order to make the estimates smoother, we may weight the observations closer to where we want our estimate than those farther away. For instance, if we want to estimate the mean travel time index at 1992, we might weight the observed value at 1992 the most and the values at 1990 and 1994 the least. As we will see, this can lead to a smooth curve similar to the smoothing spline in Figure A.2. Finally, we can choose the width of our local neighborhood to produce the level of smoothness that we want. This issue is often known as choosing the bandwidth or smoothness parameter.

Specifically, let $x_1,\ldots,x_n$ be independent and identically distributed with density (pdf) $f(\mathbf{x})$. A typical kernel estimate of $f(\mathbf{x})$ is

$$\hat{f}(\mathbf{x})=\frac{1}{s(nh)}\sum_{i=1}^{n}k\{h^{-1}s^{-1}(x-x_i)\}$$

where the function k is given by $k(x^2)=K(x)$, K is the kernel, s^2 is the sample variance, and h is the bandwidth. In the case of the example above, using moving averages, the kernel K would be the uniform density equal to 1/2 on the interval $[-1, 1]$. In all our real examples, K is the normal density.

$$K(x)=\frac{1}{\sqrt{2\pi}}e^{-x^2/2}.$$

The choice of bandwidth, h, can be complicated; see Siminoff (1996). JMP has a default choice and allows the user to override the default. For many purposes the choice of bandwidth for density estimation given by $h_{opt}=\{4/3\}^{1/5}\,\sigma n^{-1/5}$ is reasonable and is known as the normal (Gaussian) plug-in estimate of bandwidth. It would be optimal in large samples if the true density were normal. An example of JMP's kernel density estimator was given in Figure A.1.

The kernel density estimate of a regression function is given by the formula

$$r(x)=\hat{E}[Y\,|\,X=x]=\frac{\sum_{i=1}^{n}y_iK\left(\dfrac{x_i-x}{h}\right)}{\sum_{i=1}^{n}K\left(\dfrac{x_i-x}{h}\right)}$$

where K is the kernel, as previously defined. Here the plug-in bandwidth can be chosen by cross-validation (see Simonoff, 1996) or by trial and error. An example of a smoothed fit for Houston traffic data using a Gaussian kernel is shown in Figure A.2.

The JMP software uses smoothing splines to estimate regression curves using a soft modeling approach. The JMP software allows the user a trial-and-error approach using a slider. For all choices of stiffness asymptotically (related to kernel bandwidth) the user gets an R-square statistic and a sum of squares due to the error for the smoothed fit.

Another popular smoothing method is loess; see Cleveland (1979) and Cleveland and Devlin (1988). This approach is very useful when modeling nonlinear functions. The choice of smoothing parameter is an important topic. It might be wise for someone new to smoothing to ask a statistician for help with choosing an optimal smoother. Simonoff (1996) and Silverman (1986) spend a considerable amount of time on calculations needed for choosing the smoothing parameter.

REFERENCES

Cleveland, W. S. 1979. Robust locally weighted regression and smoothing scatterplots. *Journal of the American Statistical Association* 74:829–36.

Cleveland, W. S., and S. J. Devlin. 1988. Locally weighted regression: An approach to regression analysis by local fitting. *Journal of the American Statistical Association* 83:596–610.

Silverman, B. W. 1986. *Density estimation for statistics and data analysis*. London: Chapman & Hall.

Siminoff, J. 1996. *Smoothing methods in statistics*. New York: Springer.

Index